INVENTING THE FUTURE

INVENTING THE FUTURE

DAVID SUZUKI

Stoddart

First published in 1989 by
Stoddart Publishing Co. Limited
34 Lesmill Road
Toronto, Canada
M3B 2T6

CANADIAN CATALOGUING IN PUBLICATION DATA

Suzuki, David, 1936-
    Inventing the future: reflections on science, technology and nature

ISBN 0-7737-2354-4

1. Science.   I. Title.

Q171.S89 1989   500   C89-094418-0

These articles first appeared in *The Toronto Star* or *The Globe and Mail.*

Typesetting: Tony Gordon Ltd.

Printed and bound in the United States of America

To all aboriginal people
who are fighting for their identity
and their culture:

Those of us who seek to live in balance
with nature and to leave some of the
wondrous diversity of life for generations to come
have much to learn from you.

Contents

PREFACE XI

THE KNOWLEDGE OF GOOD AND EVIL
Introduction   3
Science's Jigsaw Puzzle   5
Are There No Limits?   7
The Pain of Animals   12

THE GREAT CODE: GENETICS AND SOCIETY
Introduction   21
Through Different Eyes   22
Genetics After Auschwitz   24
The Temptation to Tamper   35
Nature-Nurture Debate Lives On   37
There Will Always Be Rushtons   40
Our Fragile Democracy   43
Racing to Reach the Genome   45
Final Dance on Racism's Grave   50

TECHNOLOGY'S DOUBLE-EDGED SWORD
Introduction   55
It Always Costs   56
The Illusory Oil Change   58
Nuclear Menus (Or, Eating in the Nuclear Age)   61
PCBs as Faustian Bargain   64
The Lesson of Japan   69
The Prostitution of Academia   73
A Bid for Scientific Excellence   78

SCIENCE AND THE MILITARY: AN UNHOLY ALLIANCE
Introduction   87
Defensive Biology   88
A Technology Out of Control   90
The Ultimate Technological Fix   93
Ethnic Weapons   97
Is It Dead?   99

WARRING SIBLINGS: ECONOMICS AND ECOLOGY
Introduction   105
The New Math   106
The Ecosystem As Capital   109
On Our Own   111
Ecologists and Economists Unite!   113
Industrial Doublespeak   116
Infoglut and Its Consequences   118

THE ENVIRONMENT: THE SCOPE OF THE PROBLEM
Introduction   127
What Is the Problem?   128
It Takes a Disaster   130
Going Down Under   132
The Rape of the Amazon   135
Wilderness to the Axe   139
Trouble in the Forest   142
The Really Real   144

Rx FOR A SICK PLANET
Introduction   153
Red Herrings Galore   154
The Power of Diversity   156
Owning Up to Our Ignorance   159
Stuck in Red Tape   163
Borrowing from Children   166
Politicians Weak in Scientific Savvy   175
Making Waves   178

WONDER LOST: HOW EDUCATORS HAVE FAILED
Introduction 183
Teaching the Wrong Lessons 185
Where Are All the Women? 187
Losing Interest in Science 190
God in the Machine 194
Mad Doctors, Mad Machines 196
Why Johnny and Mary Can't Experiment 204

WONDER REGAINED: WHAT EDUCATORS CAN DO
Introduction 211
The Right Stuff 212
The System and the Ecosystem 214
It Starts in Kindergarten 216
A Gifted Teacher in Japan 218
A Love Story 220
We Are Part of the Web 223

THE ABORIGINAL WORLDVIEW
Introduction 229
Children of the Earth 230
Showdown in Brazil 234
The Invisible Civilization 237
What the Innu Know 241
Desecrating the Cathedral 243
The Spiritual Value of Land 245

Preface

*F*or two years, I wrote a weekly newspaper column for *The Toronto Star* and found, to my amazement, that I was not only able to meet the weekly deadline, but also that there was no dearth of topics to cover. I then took a few months to consider where my writing should go and decided to continue as a columnist with *The Globe and Mail,* which is available across Canada. I wrote for almost two and a half years when *The Globe and Mail* fired me for becoming too environmentally restricted.

As the host of programs on television, "Science Magazine" and "The Nature of Things," and radio, "Quirks and Quarks" and "Discovery," I am just a small part of a team of people who are involved in producing reports, from deciding on topics to the presentation and final packaging. In contrast, the newspaper columns have covered topics that are entirely my choice and the ideas contained therein reflect my priorities, thoughts and opinions.

Each article encapsulates one theme or idea, and is often prompted by an immediate news item such as a fire that spreads PCBs or an oil spill. The limitation of about a thousand words a column imposes severe constraints on the scope of each self-contained topic; however, just as a single cell of an organism may represent a complete entity but when combined with other cells becomes a part of another level of complexity from which another structure emerges, so it is with weekly columns. Each must stand on its own, but when arranged deliberately, they fit together and provide another collective perspective that cannot be covered by a single article. That is what I have tried to do in combining and organizing essays into specific topics such as genetics and education.

Today we are inundated with books about science and technology.

Most fall into two general categories: those that depict heroic epics of discovery by intellectual giants or those that emphasize the destructive consequences of scientific discovery. It is true that I tend towards the latter category, but perhaps I have an unusual perspective that is based on a different conjunction of experiences.

As a child, I was a direct victim of the effects of the braggadocio of geneticists early in this century. Of course, I only realized this long after I myself had become a geneticist. After Pearl Harbor, my family was stripped of all rights of Canadian citizenship for sharing genes that had come from the country of the enemy near the turn of the century.

Later as a university faculty member, I discovered through reading that leading geneticists early in this century had made exuberant extrapolations from studies on guinea pigs and fruit flies to behaviour of races and social groups. And those grand claims had ultimately led to restrictive immigration laws and proscriptions on interracial marriage in the U.S., incarceration of Japanese Canadians and Japanese Americans and to the horrors of the Nazi Holocaust in Germany. The recognition that scientific pronouncements have profound social consequences eventually led me to look at science from a perspective that few of my colleagues share.

That began my personal odyssey, first through the implications of genetics on society and then gradually to considerations of the nature of scientific discovery, the limits to scientific knowledge and finally to the powerful impact of scientific invention on the entire planet. I had to confront the terrible paradox — through science we now have longevity, health, comfort and material wealth unmatched in any other period of human history, yet we have not reached a utopia. Instead we are also beset with immense problems of species extinction, atmospheric degradation and global pollution that are exacerbated by the application of scientific innovation. On reflection, I think it is clear that while science provides insights into the complexity of the world around us, those insights when combined present a fractured mosaic rather than a seamless whole. There are profound limits to science that must be recognized if we are to minimize the destructive consequences of using the powers provided by scientific discovery. This book is part of an ongoing search for answers and new perspectives.

In trying to provide a different way of looking at society and an examination of our values and assumptions, I hope we may see with greater clarity. We live with numerous "sacred truths" in our belief and value system that are never questioned. For example, most of us take it for granted that nature is there for us to use in any way we can think of and that growth in consumption and wealth is a measure of "progress." It's time to re-examine some of these beliefs because they may be the cause of many of our major global problems today.

The key reason for human survival as a species was our large, complex brain. The mind that emerged from that brain conceived of a *future* as no other animals have. By inventing the future, we also gained a unique consequence — *choice*. By projecting the effect of our actions into the future, we could see there were different options that led to different consequences. So we deliberately chose the option that we thought maximized our survival.

Today, with all the amplified brainpower conferred by computers, telecommunications, scientists and engineers, we seem incapable of choosing the best option to avoid catastrophic consequences. In large part, we are blinded by deeply held beliefs and values, while all around us the signs of environmental degradation are legion. We must reinvent a future free of blinders so that we can choose from real options.

I am an optimist who writes from the perspective of a scientist, victim and parent. Although much of what I see is bleak and frightening, I believe in the power of reason to alter human behaviour. I am spurred on by the enormous shifts that have happened within a very short time — attitudes towards smoking, exercise and diet, for example. In only a few months there has been an incredible swing in attitude towards the environment, and I believe that parents are now realizing that there are different ways to think about leaving our children "better off." It is in a spirit of hope and optimism that I invite you to join me on this search for another path to the future.

THE KNOWLEDGE
OF GOOD AND EVIL

Introduction

*T*his century has witnessed two unprecedented changes on this earth. Human populations have exploded with sudden speed. It took all of the 800,000 years of human existence to reach a quarter of a billion people at the time of Christ's birth. Then it took another seventeen centuries to double to reach half a billion. We finally hit the first billion about 1830, and then with blinding speed in less than 150 years, populations had doubled twice to reach four billion. We're at five billion now and appear on our way to ten within fewer than another fifty years. We are now the most ubiquitous and numerous large mammal on earth.

The explosive increase in human numbers has been accompanied by an even steeper rise in brute muscle power conferred by science and technology. The entire history of automobiles, airplanes, antibiotics, oral contraception, nuclear energy, computers, plastics, satellites and xerography is encompassed by the span of a single human life. We can't go anywhere on the planet without using or encountering the debris of science and technology.

The huge increase in material wealth and consumption since the end of the Second World War and the accumulation of powerful methods of extraction of natural resources have generated a sense that science and technology supply the knowledge required to control and manage the entire planet. It is a terrible delusion that is not supported by an understanding of what science provides and the nature of technological power.

Another uncomfortable fact is that the vast majority of scientists and engineers in the world carry out work for the military. Such work may be called "defence research" but ultimately it translates into weapons for killing. After all, the horribly imaginative weapons — neutron, particle beam, chemical, biological, ethnic — don't come from the minds of politicians or military strategists but are the products of the fertile imagination of scientists and engineers.

The majority of the remainder of the scientific and engineering professions works for private industry — for profit. If the primary motivations underlying the application of scientific knowledge today are destruction and profit, it's hardly surprising that the best interests of the general public or long-term environmental effects seldom weigh heavily in determining whether a new discovery will be used.

In the past, the time between scientific discovery and its application might be measured in generations, so there was time to consider possible implications of using such insight. Today, the gap between discovery and application has grown astonishingly small.

As long as scientists and engineers are working for the military and industry, the general public will not have an opportunity to consider new technologies and have a say in whether or even how they will be used.

In a democracy, the people we elect are supposed to look after our interests, presumably to provide a balance to profit and power. The almost total absence of scientific and technological topics as election issues shows that, in fact, we do not expect our elected representatives to be informed or take such matters seriously.

We get the governments we deserve, so we have to be more rigorous in the standards we use to select the people we vote for.

Science's Jigsaw Puzzle

Science provides a unique way of describing the world around us. Its basic methodology is to focus on one aspect of nature, isolating it from all else, controlling everything impinging on it and measuring everything that happens. Science has enabled us to look at nature and gain insights, but ones that are fragmented into bits and pieces.

The apparent success of science in acquiring knowledge necessary to control the world is based on the assumption that if nature is reduced to its most basic parts, a comprehensive explanation will eventually be obtained by simply fitting the components back together like pieces in a giant jigsaw puzzle.

Twentieth-century physicists discovered that this hope simply doesn't hold up in the real world. They found that we cannot know anything with absolute certainty even at the most elementary level of subatomic particles. This is because the behaviour of those particles can be described only statistically, not with certainty. Physicists have also learned that the very act of observing nature changes it, because in order to see it, we have to control it; thus, we can *never* know nature as it really is.

Perhaps the most shocking discovery was that when we reduce nature to its most basic components, the description we obtain is of little value in predicting the properties of those elements when they are *combined*. In other words, though physicists can provide a detailed description of atomic hydrogen and atomic oxygen, for example, the behaviour of two atoms of hydrogen combined with one of oxygen in a molecule of water still cannot be predicted. In combination, atoms exhibit characteristics that are more than the sum of their individual properties.

As long as scientists look at nature in isolated bits and pieces, they can never describe the whole. Physicists have understood that, but by and large, biologists have not. In part that's because in the past there was a widely held notion in biology called *vitalism*. It was the belief that living things possessed a kind of vital force or spiritual essence that non-living things lack. Today vitalism is a discredited idea and anything smacking of it is rejected.

In 1962, as a new assistant professor, I once suggested in a class that the properties of a whole cell might be more than the sum of the properties of its parts. A graduate student leaped up and said accusingly, "You're a vitalist!" Years later, an eminent biochemist at the University of British Columbia scoffed at my suggestion. "The very reason we break cells open to study their parts is to understand the properties of whole cells," he told me in admonishment.

In 1986, I gained a new perspective on this issue when I interviewed Roger Sperry, a neurobiologist from the California Institute of Technology. Dr. Sperry has carried out some of the most exquisite studies ever performed on the function of neurons. His most famous discovery was that when the connection between the two hemispheres of the brain of an epileptic patient is severed, the hemispheres function independently as very different minds. He won a Nobel Prize for his work.

Neurobiologists operate on the notion that if they can obtain a comprehensive description of all the neural connections and their chemical signals in the brain, we will have a complete description of thought and behaviour. It is assumed the mind is simply the expression of the sum of the brain's components and, in principle, understandable through the study of its parts. If this were true, then there would be no such thing as free will, because our behaviour and thought would be the expressions of underlying cellular connections and chemistry, and thus completely predictable.

That belief struck Sperry as a basic contradiction of what we know at the level of matter itself. There must be, at *every* level of increasing complexity, *emergent properties* that are the result of the interaction of the constituent parts, he reasoned. On the basis of physical principles alone, there is no need to suggest some vital force.

But this obvious insight has not penetrated biology. Biochemists simply assume that by determining all the properties of subcellular components and giant molecules we will understand whole cells. Geneticists continue to believe that a complete description of human DNA (deoxyribonucleic acid) will yield a comprehensive picture of what we are. Neurobiologists have faith that human thought, imagination and emotion reside in the wiring diagrams of the brain's neurocircuits. And ecologists act on the assumption that by dissecting ecosystems into their components and modelling their interaction, we will understand the entire system. However, as Sperry points out, this faith flies in the face of what we know at the atomic level of all matter.

Sperry abandoned the area of research for which he had earned the Nobel Prize, and has spent the past fifteen years trying to determine whether it is possible to study emergent properties of complex systems in a different but still scientific way.

And for the past fifteen years, his colleagues have treated him as if he has lost his marbles. It is very sad that someone who has proved his scientific ability many times over and decided that he wanted to look at larger questions is so readily written off by his tunnel-visioned colleagues. What he is saying poses far-reaching implications about what we learn from science.

Are There No Limits?

*B*ecause most biologists and medical scientists continue to operate on notions of reductionism — reducing parts of nature to their most elementary components — concepts and insights gained in the fragmented scientific way are often applied for limited purposes. Yet they may have profound moral and philosophical repercussions that ripple far beyond the restricted view of the scientist. Nowhere is this more apparent and difficult than in our ability to intervene in the fate of newborn infants.

In 1986, Baby Andrew, an eight-month-old boy in Ontario, was suffering from a rare form of blood cancer called myelomonocytic leukemia. His case focussed interest on the question of where medical practices should stop.

During the 1950s, when I was a teenager, leukemia was a death sentence; and even today, all victims of the disease who are not treated soon die. But now the statistics on patients with its most common form, acute lymphoblastic leukemia, are astounding — with a regimen of radiation and chemotherapy, ninety per cent have remissions, and more than half remain free of the disease five years later and are considered cured. Medical science richly deserves the accolades given it for many achievements of this century in improving the quality of our lives and the average lifespan we can expect.

Baby Andrew's form of leukemia is less tractable to treatment, though up to three-quarters of the children with it who are treated may go into temporary remission, with perhaps a third of remissions lasting beyond a year or two. But treatment of children younger than a year old is far more difficult. They are usually treated more "aggressively," yet their survival prospects are very poor.

Untreated leukemia patients do not suffer intense pain or prolonged weakness — they usually die from an inability to fight off an infection or from bleeding in the brain. Treatment involves the use of chemicals that kill rapidly dividing cells. In this way, blood-producing bone marrow is wiped out in the hopes of destroying the cancerous cells. However, rapidly dividing normal cells, most notably those of the gastrointestinal tract, are also affected. Thus, chemotherapy is accompanied by intense nausea and vomiting. The treatment creates a discomfort far more excessive than that of the disease itself. It takes four to six weeks for a doctor to determine whether the patient is responding to treatment. Overall, the prognosis was that at the very best, Baby Andrew had a one-in-four chance of long-term survival.

I have deliberately stuck to the statistical and medical facts, but they serve to hide the terrible human dilemma of parents faced with the reality of a sick child, literally deathly ill at the very onset of life. Baby Andrew's parents decided the odds of his survival after treatment simply weren't worth the certain suffering that would be produced by the treatment. They chose, therefore, to refuse medical intervention, thereby assuring his early death. Child welfare authorities felt otherwise and took the case to the courts, where they succeeded in forcing the medical treatment of Baby Andrew.

The high degree of success of leukemia treatment today is built on generations of parents who opted to allow doctors to try anything in the minuscule hope of some medical miracle. Those dying children were guinea pigs whose parents willingly let them become ciphers in accumulated data. And many of those children, though already assured of dying early, had their last weeks and months made all the more cruelly painful by the experiments of medical scientists. As a parent, I hope I'll never be faced with that kind of terrible choice.

But the question still nags — how far do we go? I don't know the current success rate of organ transplants, for example, but the first ones all had to end in failure, however carefully experiments were first done with animals. Perhaps there is a quality of life gained for liver or heart recipients that justifies the operation in young children today. But what was the rationale for the experience of Baby Fae, the infant recipient of a baboon's heart? Her brief life was a grotesque media sensation that had no scientific or human justification. Will we soon transplant a miniaturized version of the Jarvik mechanical heart into younger and younger "patients"? Will we continue to create more children like David, who was imprisoned in a germ-free plastic bubble in the hope that there will be a breakthrough in dealing with his immune-deficient condition sometime in the future?

Somehow it seems to me that we have forgotten to ask a most important question — are there no limits? Has death become such an unacceptable prospect that we will pull out all stops to stave it off? In the name of medical "progress," do we demand that everyone submit to medical intervention on the chance that a statistically significant prolongation of life may ensue? The imperative to fight death at all costs cannot stem from some profound commitment to the sanctity of life. If it did, doctors could not possibly support the current levels of abortion in major cities of up to half of all pregnancies for reasons that have nothing to do with the health of the woman or the fetus.

In an increasingly secular age, science seems to have cut us loose from any sense of place and meaning. Biological science indicates that life arose on this planet by chance, that we have the form and shape that we do because of the vagaries of environment and natural selection. Bereft of meaning and alone in the universe, we find death all the more terrifying. So we choose to fight it, and each time death occurs we must admit to defeat by the forces of nature. But in fostering the notion that through science and research human beings can leap the boundaries of nature and free ourselves from the dictates of our own biology, we have created something even more frightening than death — massive medical intervention to prolong the process of dying. And it is an innocent infant with no fear of death who makes us face up to this terrible thought.

At the other end of the cycle of life, medical science and technology have derived powerful tools to intervene in the dying process, thereby rendering a definition of death much more difficult. Prolongation of life for its own sake has thrust us into the uncomfortable position of trying to play God.

On December 1, 1986, my twin sister's eldest child, Janice, fell into a deep coma from which she never regained consciousness. She was twenty-seven years old. Viral encephalitis swept through her brain, causing massive and irreversible damage. For four months, my sister and her family sat in vigil as medicine's arsenal of tools — tranquillizers to reduce the convulsions, antibiotics to limit bronchial infection, intravenous feeds to sustain nutrition and fluids — kept Janice "alive." I was astounded by the tenacity of the evolutionary mechanisms operating in Janice's body to maintain life.

Three years earlier, returning home from a movie with my father, my mother suffered a massive heart attack. She was seventy-four and for years had been showing the progressive memory loss of Alzheimer's disease. Fifteen minutes after she collapsed, she was resuscitated by paramedics, rushed to a hospital and put on a respirator in intensive care. Oxygen deprivation had permanently damaged her already deteriorating brain, but as with my niece, all Mom's survival mechanisms had kicked into action and she "lived" for another week.

It is one of the ironies of the twentieth century that the great success of modern medicine has made death so much more ambiguous and, in many cases, agonizing for the patient and family. A century ago, my mother and niece would have had rapid and humane deaths. Don't get me wrong. I wish my mother and niece were still alive and healthy and I am grateful for the emergency centres and extraordinary techniques and expertise now available in crisis circumstances. But is it monstrous to hope for death when the medically managed process of dying seems driven only to stave off at all costs an inevitable death? Often those costs are increased pain without a corresponding improvement in the *quality* of life that is gained.

We should not forget that no doctor has ever "conquered" death; at best, death is postponed. Like all other life forms, each of us will die, and no amount of scientific research and heroic measures will ever change that. What we are accumulating rapidly is an ability to intervene and disrupt the natural course of events that (still) lead to unavoidable death.

Today medicine is a costly and complicated business. We have dramatic surgical techniques, diagnostic machines and an array of powerful drugs. But in a society in which nutrition and sanitation have virtually wiped out major killers of the past, new technology itself drives doctors to the treatment of ever more rare and exotic conditions. Where once premature babies weighing less than two kilograms had a poor chance for survival, today they routinely make it. But that means that a one-kilogram baby becomes a potential target for intervention. Once-lethal congenital defects are being corrected by radical surgery in newborn infants and even in some cases in fetuses still within the mothers' bodies. At the other end of life, has death become so terrifying that we'd prefer to be tethered by a two-metre tube to a machine that keeps a mechanical heart pumping than to die of heart disease?

Once medical science has made new techniques available, we can't deny or revoke them. But when I read heart-rending stories about children who die before they can receive a third or fourth liver transplant, I feel relief for the child.

By intervening in the process of death, we create novel situations where there are often no biological models to guide us — essentially, doctors create a new kind of human being. A simple example is the development of a shunt to allow the draining of the fluid that often builds up in the skulls of children with spina bifida. Developed by an engineer who was horrified by the cranial swelling in his spina bifida child, that invention eliminated much of the brain damage and the disfiguring build-up of fluid in the skull. Children who before the shunt almost always died now survived.

Only then was it found that spina bifida is not a simple condition. The outlook for a child's life with the shunt varied from child to child. Based on experience, doctors found that the spinal lesion's position, size and severity were predictive indicators of the efficacy of installing the shunt. So today in many British hospitals doctors deliberately assess spina bifida infants at birth and place them into different categories. Some receive the shunt and every effort to provide a high quality of life, while others are allowed to die. It was technology itself that created this sorrowful dilemma.

There are many who do not want to face up to the decisions that now have to be made. Aiming for life at all costs, regardless of its quality, simply does not take into account the reality of today's technologically sophisticated world where issues are far more profound and difficult than they once were.

But medical doctors and all those who have lived with the agony of loved ones suspended in that technological twilight between life and death cannot avoid the question.

The Pain of Animals

*M*edical technology has taken us beyond the normal barriers of life and death and thereby created unprecedented choices in *human* lives. Until recently, we have taken for granted our right to use other species in any way we see fit. Food, clothing, muscle power have been a few of the benefits we've derived from this exploitation. This tradition has continued into scientific research where animals are studied and "sacrificed" for human benefit. Now serious questions are being asked about our right to do this.

Modern biological research is based on a shared evolutionary history of organisms that enables us to extrapolate from one organism to another. Thus, most fundamental concepts in heredity were first shown in fruit flies, molecular genetics began using bacteria and viruses and much of physiology and psychology has been based on studies in mice and rats. But today, as extinction rates have multiplied as a result of human activity, we have begun to ask what right we have to use all other animate forms simply to increase human knowledge or for profit or entertainment. Underlying the "animal rights" movement is the troubling question of where we fit in the rest of the natural world.

When I was young, one of my prized possessions was a BB gun. Dad taught me how to use it safely and I spent many hours wandering through the woods in search of prey. It's not easy to get close enough to a wild animal to kill it with a BB gun, but I did hit a few pigeons and starlings. I ate everything I shot. Then as a teenager, I graduated to a .22 rifle and with it, I killed rabbits and even shot a pheasant once.

One year I saw an ad for a metal slingshot in a comic book. I ordered it, and when it arrived, I practised for weeks shooting marbles at a target. I got to be a pretty good shot and decided to go after something live. Off I went to the woods and soon spotted a squirrel minding its own business doing whatever squirrels do. I gave chase and began peppering marbles at it until finally it jumped onto a tree, ran to the top and found itself trapped. I kept blasting away and grazed it a couple of times so it was only a matter of time before I would knock it down. Suddenly, the squirrel began to cry — a piercing shriek of terror and anguish. That animal's wail shook me to the core and I was over-whelmed with horror and shame at what I was doing — for no other reason than conceit with my prowess with a slingshot, I was going to *kill* another being. I threw away the slingshot and my guns and have never hunted again.

All my life, I have been an avid fisherman. Fish have always been the main source of meat protein in my family, and I have never considered fishing a sport. But there is no denying that it is exciting to reel in a struggling fish. We call it "playing" the fish, as if the wild animal's desperate struggle for survival is some kind of game.

I did "pleasure-fish" once while filming for a television report on the science of fly fishing. We fished a famous trout stream in the Catskill Mountains of New York state where all fish had to be caught and released. The fish I caught had mouths gouged and pocked by previous encounters with hooks. I found no pleasure in it because to me fish are to be caught for consumption. Today, I continue to fish for food, but I do so with a profound awareness that I am a predator of animals possessing well-developed nervous systems that detect pain. Fishing and hunting have forced me to confront the way we exploit other animals.

I studied the genetics of fruit flies for twenty-five years and during that time probably raised and killed tens of millions of them without a thought. In the early seventies, my lab discovered a series of mutations affecting behaviour of flies, and this find led us into an investigation of nerves and muscles. I applied for and received research funds to study behaviour in flies on the basis of the *similarity* of their neuro-muscular systems to ours. In fact, psychologists and neurobiologists analyse behaviour, physiology and neuroanatomy of guinea pigs, rats, mice and other animals as *models* for human behaviour. So our nervous systems must closely resemble those of other mammals.

These personal anecdotes raise uncomfortable questions. What gives us the right to exploit other living organisms as we see fit? How do we know that these other creatures don't feel pain or anguish just as we do? Perhaps there's no problem with fruit flies, but where do we draw the line? I used to rationalize angling because fish are cold-blooded, as if warm-bloodedness indicates some kind of demarcation of brain development or greater sensitivity to pain. But anyone who has watched a fish's frantic fight to escape knows that it exhibits all the manifestations of pain and fear.

I've been thinking about these questions again after spending a weekend in the Queen Charlotte Islands watching grey whales close up. The majesty and freedom of these magnificent mammals contrasted strikingly with the appearance of whales imprisoned in aquariums. Currently, the Vancouver Public Aquarium is building a bigger pool for some of its whales. In a radio interview, an aquarium representative was asked whether even the biggest pool can be adequate for animals that normally have the entire ocean to rove. Part of her answer was that if we watched porpoises in the pool, we'd see that "they are quite happy."

That woman was projecting human perceptions and emotions on the porpoises. Our ability to empathize with other people and living things is one of our endearing qualities. Just watch someone with a beloved pet, an avid gardener with plants or, for that matter, even an owner of a new car and you will see how readily we can personalize and identify with another living organism or an object. But are we justified in our inferences about captive animals in their cages?

Most wild animals have evolved with a built-in need to move freely over vast distances, fly in the air or swim through the ocean. Can a wild animal imprisoned in a small cage or pool, removed from its habitat and forced to conform to the impositions of our demands, ever be considered "happy"?

Animal rights activists are questioning our right to exploit animals, especially in scientific research. Scientists are understandably defensive, especially after labs have been broken into, experiments ruined and animals "liberated." But just as I have had to question my hunting and fishing, scientists cannot avoid confronting the issues raised, especially in relation to our closest relatives, the primates.

People love to watch monkeys in a circus or zoo and a great deal of the amusement comes from the recognition of ourselves in them. But our relationship with them is closer than just superficial similarities. When doctors at Loma Linda hospital in California implanted the heart of a baboon into the chest of Baby Fae, they were exploiting our close *biological* relationship.

Any reports on experimentation with familiar mammals like cats and dogs are sure to raise alarm among the lay public. But the use of primates is most controversial. In September 1987, at the Wildlife Film Festival in Bath, England, I watched a film shot on December 7, 1986, by a group of animal liberationists who had broken into SEMA, a biomedical research facility in Maryland. It was such a horrifying document that many in the audience rushed out after a few minutes. There were many scenes that I could not watch. As the intruders entered the facility, the camera followed to peer past cage doors, opened to reveal the animals inside. I am not ashamed to admit that I wept as baby monkeys deprived of any contact with other animals seized the fingers of their liberators and clung to them as our babies would to us. Older animals cowered in their tiny prisons, shaking from fear at the sudden appearance of people.

The famous chimpanzee expert, Jane Goodall, also screened the same film and as a result asked for permission to visit the SEMA facility. This is what she saw (*American Scientist,* November-December 1987):

Room after room was lined with small, bare cages, stacked one above the other, in which monkeys circled round and round and chimpanzees sat huddled, far gone in depression and despair.

Young chimpanzees, three or four years old, were crammed, two together into tiny cages measuring 57 cm by 57 cm and only 61 cm high. They could hardly turn around. Not yet part of any experiment, they had been confined in these cages for more than three months.

The chimps had each other for comfort, but they would not remain together for long. Once they are infected, probably with hepatitis, they will be separated and placed in another cage. And there they will remain, living in conditions of severe sensory deprivation, for the next several years. During that time they will become insane.

Goodall's horror sprang from an intimate knowledge of chimpanzees in their native habitat. There, she has learned, chimps are nothing like the captive animals that we know. In the wild, they are highly social, requiring constant interaction and physical contact. They travel long distances, and they rest in soft beds they make in the trees. Laboratory cages do not provide the conditions needed to fulfill the needs of these social, emotional and highly intelligent animals.

Ian Redmond (*BBC Wildlife,* April 1988) gives us a way to understand the horror of what lab conditions do to chimps:

> Imagine locking a two- or three-year-old child in a metal box the size of an isolette — solid walls, floor and ceiling, and a glass door that clamps shut, blotting out most external sounds — and then leaving him or her for months, the only contact, apart from feeding, being when the door swings open and masked figures reach in and take samples of blood or tissue before shoving him back and clamping the door shut again. Over the past 10 years, 94 young chimps at SEMA have endured this procedure.

Chimpanzees, along with the gorilla, are our closest relatives, sharing ninety-nine per cent of our genes. And it's that biological proximity that makes them so useful for research — we can try out experiments, study infections and test vaccines on them as models for people. And although there are only about 40,000 chimps left in the wild, compared to millions a few decades ago, the scientific demand for more has increased with the discovery of AIDS.

No chimpanzee has ever contracted AIDS, but the virus grows in them, so scientists argue that chimps will be invaluable for testing vaccines. On February 19, 1988, the National Institute of Health in the U.S. co-sponsored a meeting to discuss the use of chimpanzees in research. Dr. Maurice Hilleman, Director of the Merck Institute for Therapeutic Research, reported:

> We need more chimps . . . . The chimpanzee is certainly a threatened species and there have been bans on importing the animal into the United States and into other countries, even though . . . the chimpanzee is considered to be an agricultural pest in many parts of the world where it exists. And secondly, it's being destroyed by virtue

of environmental encroachment — that is, destroying the natural habitat. So these chimpanzees are being eliminated by virtue of their being an agricultural pest and by the fact that their habitat is being destroyed. So why not rescue them? The number of chimpanzees for AIDS research in the United States [is] somewhere in the hundreds and certainly, we need thousands.

Our capacity to rationalize our behaviour and needs is remarkable. Chimpanzees have occupied their niche over tens of millennia of biological evolution. *We* are newcomers who have encroached on *their* territory, yet by defining them as *pests* we render them expendable. As Redmond says, "The fact that the chimpanzee is our nearest zoological relative makes it perhaps the unluckiest animal on earth, because what the kinship has come to mean is that we feel free to do most of the things to a chimp that we mercifully refrain from doing to each other."

And so the impending epidemic of AIDS confronts us not only with our inhumanity to each other but to other species.

THE GREAT CODE: GENETICS AND SOCIETY

Introduction

*N*owhere is the impact of modern science on our lives and ideas of what we are as a species more profound than in the discipline of genetics, the science of heredity. Of course, I have a bias because for a quarter of a century, it was my profession.

The invention of agriculture about ten millennia ago led to our earliest and most powerful insight into heredity — *like begets like*. Thus, the selection and breeding of animals and plants with desired traits quickly led to the creation of domestic forms farmers still use today. It's not surprising that people have also wondered whether the principles of plant and animal breeding could also be applied to ensure that healthy babies were born. But only when the science of genetics was born in 1900 did we acquire the knowledge to act on this.

In this century, we have accumulated a detailed understanding of the principles that govern the transmission of traits from one generation to the next, the physical location and behaviour of "genes," the factors of heredity, and the structure of DNA, the actual molecule that genes are made of. This knowledge has been accompanied by powerful tools that enable us to manipulate the genetic material and thereby alter the genetic blueprint of organisms at will.

These manipulative capabilities hold the promise of increasing productivity, variety and utility in agriculture, forestry and pharmacology. But as novel organisms that have never existed before are created, scientists are faced with the troublesome question of unknown hazards that could result. It was the proposal by molecular biologists to insert a gene from a cancer-causing virus into a bacterial species found in the intestines of all human beings that first focussed attention on this question. But the most difficult problems arise in the area of human biology and reproduction.

Through Different Eyes

*C*hildren have such a delicious way of ignoring all social conventions to ask the obvious question. That's what happened the other day: a child stopped on the street and said to me, "Hey, mister, how can you walk with your eyes closed?"

Evolution selected the epicanthic fold of eyelids characteristic of my Mongoloid ancestors. Those fat-filled lids worked well in frigid climates, insulating eyeballs against the cold while blocking out a lot of the reflected light from ice and snow. All you need is a slit to let light into the pupil — my field of vision is just as great as the "big eyes" of the West.

But what is functional and adaptive biologically is not necessarily acceptable culturally. Small eyes are not in vogue, and Western standards of beauty are pervasive. Even in Japan today, it's not uncommon to see people walking around with one eye bandaged because they've had a cosmetic operation to put a fold into the eyelid to give the illusion of bigger eyes. (They do one eye at a time.)

I'm not immune. In the years immediately following the incarceration of Japanese Canadians as enemy aliens, my eyes were a great source of shame to me. I would have given anything to have had an eye operation. (I wanted to dye my hair and change my name too.) I'm terribly self-conscious without my glasses because I live with the illusion that the frames make my eyes look bigger. So there you have the primal essence of my hang-ups.

You may ask, "Who cares?" or "So what?" Well, I'm sure a lot of people have their own personal sensitivities to what they perceive as blemishes. When each of my children was born and I was assured they were normal, the first things I checked out were their eyes and noses. Today we live in a time when people are deliberately limiting the number of children and hence each birth has become vested with far greater importance. We want so much for our children and would like them to start out with the maximum of advantages. So if music affects the fetus, we'll play Beethoven every chance we get. What will happen as techniques for monitoring our babies increase the kinds of traits detectable and alterable before birth? If a physical feature has come to be regarded as an impediment, imaginary or not, most of us would like to avoid it in our children, if possible.

Already the crudest kind of prenatal diagnosis by amniocentesis and chorion biopsy allows a quick detection of sex. In clinics in India and China, an XX (female) chromosome constitution is not considered as desirable as XY (male), so through abortion females are being "terminated." Their "defect" is the absence of a Y chromosome. Sex selection by abortion happens here too, though doctors are understandably reluctant to discuss it. Consider a family in which three or four girls have been born. The parents would like to have a son, and if the mother is in her late thirties, she is eligible to have amniocentesis to check for Down's syndrome in the fetus. Once the test is done, parents can ask what the sex of the fetus is. If the fetus is a normal female, this family might contemplate an abortion in order to try for a boy.

The problem for society is that once the technologies become available, all kinds of unanticipated uses are thought up. Amniocentesis, the recovery of fetal fluid and cells from the amniotic sac of a pregnant woman, yields a crude biological profile of the fetus. Such information has allowed parents who carry genes for a known hereditary defect to risk pregnancy, knowing that a defect can be detected in the test and a decision can be made whether to abort. Access to the test has resulted in the birth of babies who otherwise would never have been deliberately conceived, and a great deal of anguish and suffering has been avoided. There are many terrible neurological and anatomical defects that can now be circumvented, but for many (including me) abortion of a Down's syndrome fetus takes us into a troubling "grey zone." I know families that have been devastated by the birth of a Down's child, but I also known others whose lives have been enriched enormously by a Down's infant. Once we have options, we are faced with the terrible decision where we draw the line and what we are willing to tolerate.

Now with ultrasonography, fetoscopy, chorion biopsy and restriction enzyme mapping, techniques are rapidly evolving to provide very sophisticated insights into the genetic makeup of fetuses. And if Down's syndrome raises difficulties, what of those diseases in which sound minds are accompanied by progressively worsening physical conditions — muscular dystrophy, hemophilia, sickle cell anemia, cystic fibrosis, multiple sclerosis and others? Would those who suffer from these debilitating diseases and their family members prefer that they had not been born? And as technology improves, what will we decide about scoliosis, diabetes, cleft lip and palate, albinism?

Already the U.S. government has given its approval to begin gene therapy, the deliberate introduction of DNA to alter the genetic makeup of children. Like the technologies of prenatal diagnosis, this will open up an awesome array of possibilities that many will find irresistible. I am glad that my reproductive days are over so I don't have to deal with the very real possibilities. For if the option had been available, would I have been able to resist tampering with the conformation of my children's eyes?

Genetics After Auschwitz

*T*he best guide we have to help us through the maze of ethical questions that are created by genetic engineering is history; we forget its lesson at the risk of repeating the same mistakes. Consider Josef Mengele, the infamous doctor at the Nazi death camp at Auschwitz. He was in the news again a few years ago when forensic scientists eventually concluded that bones discovered in a grave in Brazil were indeed Mengele's remains. Recalling his activities should give every scientist pause.

Mengele gained his notoriety for his experiments in genetics. I was trained as a geneticist, yet never in all the years of my education or during my entire career as a scientist did I encounter his name except in the popular press. In the field of science, Mengele does not exist.

I went to a liberal arts undergraduate school that is ranked as one of the top in North America. My first genetics professor — my inspiration and hero — was a Jew who had received his Ph.D. from Curt Stern, one of the most important figures in classical genetics. Stern was also a Jew who had fled Nazi Germany and eventually ended up at Berkeley.

I earned a bachelor of arts degree, and though I was enrolled in honours biology, the curriculum stipulated that no more than half my courses be in science. So I had wonderful courses in music, Michelangelo, twentieth-century history, world religion and literature. It was a marvellous opportunity rarely available to science students today, yet I never did encounter the philosophy or history of science. I was never taught about the excitement geneticists had felt at the turn of the century over the notion of improving the human species through selective breeding.

No one told us that geneticists had made bold claims about the "hereditary basis" for racial inferiority or superiority (which are value judgements, not terms that have any scientific meaning). Some of the leading geneticists in the twenties and thirties wrote of the genetic basis for nomadism in gypsies, criminality, drunkenness and vagrancy, but we budding scientists did not learn of the important role that our predecessors had played in encouraging laws regulating immigration from certain countries, prohibiting interracial marriage and sterilizing patients in mental institutions.

During my education in graduate school, I was never taught that being a scientist entailed enormous social responsibilities. We did not learn that the social context and value system within which scientific investigations occur affect the kind of research done and the way results are interpreted. No one told us there are limits to science, that it provides only a fragmented view of nature that can never encompass the whole. There is no code of ethics governing our activity, nor is there a sense that it is a privilege for a scientist to have public support.

Today's science student has a heavy load of science courses and little opportunity to take others outside the discipline. Science students in my university don't have room in their course loads for philosophy, history, religion or literature. Indeed, I remember one of the professors in my department during the student protests of the 1970s saying, "Okay, look. We'll keep all the best zoology courses for our honours and majors students and we'll *educate* the rest of them." Needless to say, he was sneering at the term *educate*.

Today so much is happening in science that it is difficult for a scientist to stay abreast of all developments. Even though most research papers published will turn out to be wrong, trivial or unimportant in a few years, we feel compelled to emphasize the latest work, thereby ignoring more and more of the classical studies and history. But what a loss that is.

We don't learn that geneticists were the prime movers behind the Nazi Race Purification program and that the voices of opposition to Hitler from scientists and doctors were silent. It was our colleagues, the likes of Mengele, who were carrying out their so-called research in death camps.

I know my fellow scientists find it easy to dismiss the doctors of death like Mengele. We say, "Hell, he was a medical doctor carrying out pseudoscientific experiments. He wasn't a geneticist at all, he was a nut." But "real" geneticists in Germany didn't say so. Besides, is it enough to write him off as a grotesque caricature of a scientist, a freak who happens once in a generation?

I don't think so. Germany boasted some of the most eminent geneticists and biologists of the day when the Nazi program was set in motion. The history of scientists not only in countries like Nazi Germany but in Fascist Italy, Japan and during the McCarthy years in the United States is not a proud one. As Cornell University historian Joseph Haberer has written: "What becomes evident is that scientific leaders, when faced with a choice between the imperatives of conscience and power, nationalism and internationalism, justice and patriotism, invariably gravitated toward power, nationalism and patriotism." It's a sobering indictment and one that can only be avoided in future if we remember the past.

I do not believe for a minute that Josef Mengele was merely an aberration, who can therefore be lightly dismissed. Scientists are often driven, consumed, focussed on the immediate problem at hand, and this is the great joy and strength of involvement in science. But it can also blind *any* of us to wider implications of what we and our peers are doing.

Mengele was one within a vast range of people who call themselves scientists. He was some scientist's student, he was a colleague, a peer of the medical-scientific community. It lets us off the hook too easily to say simply that he was a monster.

The explosion of the atomic bomb smashed the romantic notion of scientific innocence. The recent brouhaha over recombinant DNA and the vocal objections to the American Strategic Defense Initiative (Star Wars) are hopeful signs that the horrible silence of the scientific community may not happen again. But unless we acknowledge the likes of Josef Mengele and include him in courses taken by science students, we could quickly forget the lesson of history he provides.

The reluctance of scientists to take an unflinching look at their own history was clear when I was invited in the fall of 1987 to attend a meeting in Toronto of the organizing committee of the Couchiching Conference. Each year, experts from various disciplines meet in Couchiching, Ontario, to discuss a specific topic. In 1987 the subject was the Rise and Fall of the American Empire. In 1988 the conference was to look at DNA and genetic engineering, a timely topic in view of the tremendous advances in molecular genetics and the proposal to decipher the entire genetic blueprint of a human cell.

The Couchiching committee focussed on the technology of DNA manipulation and its future implications, which are fascinating. But the really important questions have to do with what scientists and people in power will do with the unprecedented ability to alter the genetic makeup of life forms. And the only way to anticipate that comes from looking backward.

At the Couchiching committee meeting, I reminded them that early in this century a brand-new science — genetics — had made spectacular and rapid discoveries about the laws of heredity. Scientists were understandably excited and exuberant at the potential to apply this knowledge for the benefit of humankind. I pointed out that in prewar Germany, where culture and science were at a peak, doctors and scientists had embraced the possibility of applying the benefits of genetic discovery. By extrapolating from studies of the inheritance of *physical* characteristics in fruit flies and corn plants to *behaviour* and *intelligence* in people, they concluded that human beings could be "perfected" through selective breeding and elimination of "defectives." The Nazi Race Purification programs seemed to represent the application of some of the most "progressive" ideas in science.

Thus, I suggested, doctors and scientists — especially geneticists — had, in their intoxication with new findings, popularized the notion of the overriding importance of heredity in human behaviour and sold it to Hitler's National Socialists. It led inexorably to the horrors of the Holocaust, which scientists must therefore acknowledge some responsibility for. Two of the committee members (one a molecular biologist and both Jews) were outraged. They denied the suggestion that scientists have to bear some of the blame for the excesses of Nazi action and accused me of being "hysterical."

This selective memory of science's history amounts to a coverup and a revisionism that only ensures that the same thing could happen again. Even a suggestion that there is an unpleasant aspect to science's past is interpreted as opposition to science. Three years ago, I hosted a television series; one of the programs included some history of genetics. In reviewing that show, Stephen Strauss, *The Globe and Mail*'s science writer wrote: "You are of the clergy, a scientist who left the monastery/laboratory to reform the world's understanding of his faith, and who now may well be on the way to becoming a heretic. (Some of your fellow geneticists think that)." An accusation or even suggestion of heresy powerfully reinforces dogma and threatens dissenters with excommunication from the scientific community. I can understand why scientists are reluctant to face up to the past. Fortunately, a few exceptional ones won't let us forget.

Benno Müller-Hill is a professor of molecular biology in Cologne, West Germany. Few texts in molecular biology and genetics fail to mention his work. While at Harvard University working with Walter Gilbert (who later earned a Nobel Prize), Müller-Hill carried out a classic experiment that allowed the isolation and purification of a protein molecule called a *repressor*, which controls gene activity. Only a few copies of the repressor are present in each cell, so the Müller-Hill/Gilbert experiment was a scientific tour de force. Müller-Hill's lab went on to collect sufficient quantities of the material to determine the primary structure of the protein. A scientist of world-class stature, he continues to do research. For the past decade he has also studied the history of genetics in Nazi Germany.

One of Müller-Hill's articles, entitled "Genetics After Auschwitz," appeared in *Holocaust and Genocide Studies*, volume 2, number 1 (1987). It is a document that is at once chilling and agonizing — an unflinching look by a scientist at the role of scientists in the Holocaust.

The article opens this way: "The past must be recollected and remembered before it can be evaluated . . . It is particularly difficult for scientists. Science is oriented to the present . . . only today's results exist. Only new data or new theories bring glory, honour and money for new research. Reflection on the past almost excludes the reflecting scientist from the ranks of present-day science." He goes on to summarize his studies and his book: "The rise of genetics is characterized by a gigantic process of repression of its history." Later he writes: " . . . geneticists have refused — and even now refuse — to acknowledge their history."

Science is an activity that transcends national boundaries because it is practised by an international community sharing knowledge through freely available publications. As governments around the world tie scientific innovation to their economic well-being, scientists are under tremendous pressure to do work that is socially "relevant" or "practical." Most research grants are now awarded and renewed on the basis of these considerations, and the potential economic rewards of an application of new ideas is great. Scientists focus even more than ever on work going on now and the future, but is it possible that scientists, for the most humane of reasons, can participate in work that exceeds ethical lines? If the answer is yes, then surely it behooves us to pay attention to what has happened in the past. Today most scientists are too busy, or they ignore or selectively recall their history.

When Müller-Hill went to the archives of the German Research Association in 1981 to look at extensive historical records, he was told that he was the first person to do so since the end of the war. He was seeking answers to the questions, Was Auschwitz the result of pure scientific thinking? How does scientific reason change into the greatest unreason? Is this an inevitable process of the growth of pure scientific reason? Even asking such important questions carries the risk of denial and hostility from the scientific profession.

The German biographer P. Fischer illustrates this risk in his biography of Nobel laureate and geneticist Max Delbrück. In 1947, Nobel laureate H. J. Müller, as president of the American Society of Genetics, asked Delbrück, who was returning to his native Germany, "to gather first-hand information about geneticists still in Germany, and to investigate whether any evidence exists that might absolve them from the guilt of having actively supported Nazism and prostituted genetics under the Nazis." Delbrück failed to do so because "he did not have the courage; he considered it improper to inquire which German scientists had related to the National Socialist government. So he failed to fulfill the mission entrusted to him by the American Genetic Commission . . . Later Delbrück was convinced by one of his closest friends, plant geneticist Georg Melchers, that virtually no German biologist had ever worked towards furthering the race theories, and that biology in general was not guilty of formulating the inhuman ideologies which resulted in inhuman action." Thus, even a scientist as great as Delbrück could not resist the pressures from his peer group and finally accepted the party line.

In the article "Genetics After Auschwitz" Müller-Hill says:

The number of medical doctors who guided scientific selection in Auschwitz by killing through gas or slave labour was small. Nevertheless nine university professors were actively involved in selecting 70,000 mentally ill persons for the gas chambers . . . . The exact numbers of Jews killed in Auschwitz is unknown. Two-and-a-half million is the number Höss got from Eichmann . . . . Auschwitz was not only a site of destruction and a laboratory of human biology; it was also planned as a place of chemical production. IG-Farben had placed their largest investment during the war near Auschwitz . . . . When one regards the potential for scientific investigation, production and destruction, Auschwitz turns into a monument of modern science and technology. Human biology and technical chemistry should never have been the same again after Auschwitz was liberated, but nothing changed. The human biologists and medical doctors who were not caught in the act escaped by a semantic trick; this was pseudo-science or pseudo-medicine, they said, so they were free to start again with real science and medicine.

Müller-Hill traces the web of German scientists who were connected to the work carried out at Auschwitz. It included some of the leaders of the day. The most infamous of the scientists at Auschwitz was Josef Mengele.

Mengele was not just some scientist from the provinces. He had worked with the foremost German scientists of his time: firstly with the anthropologist Professor Theodor Mollisson in Munich, then with the human geneticist and specialist in internal medicine, Count Professor Otmar von Verschauer. Mengele was not an amateur who operated outside the established structures of science.

Von Verschauer, Mengele's mentor, applied for grants for work at Auschwitz on "human twin studies, human eye defects, human tuberculosis and . . . specific serum proteins." The grants were all approved. Müller-Hill continues:

What happened to the twins? They were analysed anthropologically, physiologically and psychologically. Anything that could be measured was measured. Those who had interesting anomalies were killed by Mengele or his helpers. The interesting organs were sent to the K.W.I. for Anthropology at Berlin-Dahlem . . . .

Was Professor von Verschauer a unique case? No, other researchers had already profited from the murder of the insane . . . . Some 70,000 patients of German mental institutions were killed by gas during 1940 and the first half of 1941.

Many of their brains ended up in scientific laboratories.

What is astonishing is that to most geneticists these infamous scientists and events don't exist: they have been expunged from history.

For biologists, Nazi Germany provides important lessons that might temper our rush to exploit new ideas uncritically. But according to Benno Müller-Hill, there has been a systematic suppression and revision of the history of science under the Nazis. Müller-Hill provides an explanation of how this came about:

Auschwitz had just reached its highest destructive potential when the paper appeared which showed that DNA was the basic genetic material. It took several years until the significance of this discovery was generally understood, but when the double helix was published in 1953, only fools did not realize that genetics had virtually exploded. The speed of this development left no time for looking back or for regrets over the blood and tears that had been spilled in the process. Scientists discuss George Orwell's *Animal Farm* and *1984* and do not see that they themselves have created a universe which is equally frightening. No secret police forces them to forget the past. They obliterate it themselves on the marketplace of science. They have come to believe that they have a beautiful past, or perhaps no past at all. The chapters of textbooks which deal with genetics and society contain only a few token sentences about National Socialism.

If history is not remembered, can the scientific community be involved in horrors like those of Nazi Germany again? Of course, though undoubtedly in a different manifestation. Scientists are, above all else, human beings with all the foibles, idiosyncrasies and diversity found in any other group of people. Ambition, driving curiosity, desire for power, thirst for financial security, fear — there are many reasons why people do what they do. And in the current scramble to capitalize on the enormous potential of genetic engineering, organ transplants and a cure for AIDS, individual scientists have not been above cutting corners or compromising on ethical standards.

In part, the very methodology of science itself makes this easier, says Müller-Hill:

> Scientists observe and analyse objects. An object is a thing without rights. When a human being becomes an object he is nothing but a slave. What interests the scientist is the answer to the question he asks the object, but not the object's own questions. In general, the scientist never analyses the whole object but only a small part of it. Others dismember the object, he receives only one part of it for his analysis. The answers which he expects from the part he analyses may be numbers, DNA sequences or images . . . . This process of objectivization of the whole world, and finally of oneself as part of science, seems the main interest and pleasure of the scientist's brain. There is little place for other things in the scientist's mind.

But, it is often countered, weren't the people who carried out the atrocities in Nazi Germany second-rate intellectuals, mediocre but ambitious opportunists? Müller-Hill disagrees: "It was not in the interest of the Nazi elite that the sciences be dominated by a mob of liars and charlatans. The Nazis needed functioning science and technology to assist their wars of robbery and destruction."

But could there be a repetition of what was done by the Nazis?

The killing of deficient newborn babies as practised in Germany between 1939 and 1945 has simply become anachronistic. Most geneticists sincerely believe that here they have created new values. They do not see that they appeal to the forces of the market which state that cost-efficiency considerations make it advisable, for both parents and state, to destroy the cost-inefficient embryo.

Müller-Hill's ideas are not pleasant and he has encountered naked hostility from his scientific peers. But unless we hear him out and dig out the bad as well as the good in science's history, we will ensure that scientists will continue to do terrible things for what seem to be the highest reasons — just as their predecessors did. We need to remember that scientists are human too.

Geneticist R. Gold wrote a letter to *The Globe and Mail* (February 17, 1989 — "Gene Designers") to rebut the columns I wrote on Müller-Hill's work, yet Gold's letter itself provides an illustration of how revisionism is enforced by well-meaning scientists.

One tactic of the revisionist is to set up a straw man that can be knocked down. Thus, he writes:

> . . . in his columns, Dr. Suzuki gives intermittent indications that he wants to go further than this to argue that there is a sort of evil inherent in the science of genetics and in geneticists themselves. He seems to imply, that if we are not carefully watched, we will be up to our old tricks again and indeed may already be engaging in nefarious schemes to harm humanity.

He goes on to say that I imply "that evil occurs because scientists are prone to evil."

He infers exactly the *opposite* to what I have written throughout my career in popularizing science. I have constantly emphasized that scientists are *no better or worse* than any other group of people, but it was enthusiasm over new insights in the mechanics of heredity early in this century that led some of the finest scientists to proclaim that human beings could be "improved" through selective breeding. The goals were laudable — to avoid suffering and improve the human condition — yet those ideas were warped into Nazi race purification, again with the encouragement of some scientists. In our current excitement over the rapidly accumulating manipulative powers of DNA, many of the same claims are again being made.

Gold is a geneticist and surely knows that one of our colleagues at Harvard deliberately violated the federal guidelines on recombinant DNA experiments and, when discovered, quit the university to set up his own private biotechnology company. It was a UCLA doctor who violated federal grant restrictions by administering DNA to children with hereditary diseases. A Montana professor went ahead with prohibited field studies of DNA-engineered organisms in elm trees. There are other deliberate violations of regulations by reputable scientists. It is precisely because they are *not* evil or fanatics, but ordinary human beings who are totally caught up in their own ambitions and beliefs, that we must remember the pitfalls that the past reveals.

As I've mentioned, Müller-Hill pointed out another way that postwar German scientists rationalized what happened during the Holocaust so the events could be discounted: "This was pseudo-science or pseudo-medicine, they said, so they were free to start again with real science and medicine." That's precisely what Gold does: "The concept of racial purity is not a scientific concept and has no place in genetics. It is simply intellectual rubbish that was dragged in to justify actions undertaken for other reasons." Thus Gold rewrites history.

Early in this century, leading scientists — geneticists and anthropologists — had popularized the idea of scientifically improving the human condition by preventing "inferior" people from reproducing or by encouraging those considered "superior" to have more children. They established the social climate for the adoption of race purification policies by Nazi Germany. Seen today, these ideas are indeed "intellectual rubbish," but in the twenties and thirties they were serious scientific proposals. The important lesson is that we should be very careful about rushing to apply in society or the ecosystem ideas derived in the laboratory. But we won't learn that essential lesson if we persist in papering over the past.

Gold's final accusation is that I have not addressed "what we should do about all this. Should we stop doing the research that will continue to provide us with these choices, and, if not, what advice has he as to the choices we should make?" As a once-practising scientist, I continue to take enormous vicarious delight in the insights gained by scientists and have long written of the need for better support of good science in this country. But it would be foolhardy to suggest that there are no detrimental or unpredictable negative consequences of the application of science.

Gold seems unwilling to face up to the vast changes that have happened in science and its relationship to society in a few decades. In half a century, human population has more than *doubled*, while the per capita consumption of resources in the highly industrialized nations has increased many times more. In the same period, the scientific community has greatly expanded its numbers, just as the interval between discovery and application has decreased radically. Surely then, there must be constant reassessment of the changing relationship between science and society, and the best guide that we have is history. Accusations of heresy, neo-Ludditism or anti-intellectualism become powerful means of discouraging critical dissent or enquiry. In the long run, that does an enormous disservice to the public *and* to science itself.

The Temptation to Tamper

*H*aving looked at the lessons from the past, we can now see how we may be tempted to ignore them and repeat those actions. The temptation arises over the question of defining what is normal. Or perhaps it should be asked, what is abnormal? That is the troubling issue raised every time we develop technologies that hold out the possibility of overcoming defects. For example, sterility in some men and women has existed since time immemorial. It is not a life-threatening condition, but once artificial insemination and embryo transplants became possible, sterility became a medical problem. Indeed, a lot of medicine is now concerned with treating desires (to have a child, smooth skin or larger breasts), thereby converting normal human variability into abnormalities or defects.

What happens when we explore the possibilities that flow from genetic engineering? Biotechnology is the hot research area right now. Through molecular biology, we can engineer microorganisms to do our bidding. Many molecules normally made by the human body can now be manufactured in large quantities by bacteria.

One of the most spectacular feats in this area was the synthesis by the American biotechnology company Genentech of the gene specifying human growth hormone (HGH). The gene was inserted into bacteria, which then proceeded to turn out a protein identical to HGH. This hormone is normally produced in the pituitary gland, a small lobe of the brain.

HGH is an important growth regulator, and when it is missing or present in low amounts, pituitary dwarfism results. In the past, HGH was extracted from the pituitary glands of cadavers and used to stimulate growth in youngsters whose height at maturity was projected to be well below average. In the United States, the National Hormone and Pituitary Program processed up to 50,000 glands a year to supply 3,500 children and adolescents. HGH was in limited supply and expensive. Genentech's achievement has meant that there is a limitless supply of HGH and it should therefore become relatively cheap.

But now supply far exceeds demand. So in fine free-enterprise style, there has been a search for new potential markets for this product. Very little is known about the subtle biological effects of HGH. We know it is abnormally low in pituitary dwarfs, but we don't really know how well correlated the amount is with the height of the population generally. It does appear that HGH induces the loss of fat without a loss of muscle. On that basis, HGH is being proposed as a treatment for obesity. Meanwhile, athletes, ever watchful for any possible advantage in their competitive sports, are already taking the drug to induce preferential buildup of muscle mass. I can't overemphasize the extent of our ignorance about the normal metabolic role of HGH.

But the greatest potential to market plentiful HGH lies in the redefinition of what is abnormally short stature. Dwarfs lacking HGH are clearly defined. But in a distribution curve of normal people, there will always be those who fall at the short end of the height spectrum. At what point does a normally short person classify as "abnormal"? Apparently it's all in the mind. A UCLA professor described a father who brought in his son to be tested for HGH deficiency. When told that there was no cause for concern because all indications were that the boy would grow to a height of five feet seven inches, the father roared, "That's absolutely unacceptable!" and demanded HGH for his son.

Ours is a society in which size is connected to social rewards. Executives and bank presidents are taller than average, while bishops exceed priests in average height. Certainly in North America the rewards of size on the athletic field are staggering. The take-home pay of a towering basketball star or a football behemoth is immense. Any sports fan knows how fanatically some parents pursue a career for a child. It is not at all wild speculation to suppose that ambitious parents will line up to purchase HGH for their children. They will willingly put their children at unknown risk of side effects of the treatment. Doctors already report requests for HGH "therapy."

What we are talking about now are perfectly normal, healthy children who are going to be treated as if they are somehow defective and in need of medical treatment. A report last fall in the journal *Science* suggests that "it is almost certain that many affluent parents of short children will have their children treated with human growth hormone as a matter of course."

Indeed, James Tanner of the University of London predicts that HGH treatment may become as accepted as orthodontics. Some scientists suggest that there is a need to do more experiments, to find out what the effects of HGH therapy might be. They speak of the need for placebos and double-blind experiments. But why? There ought to be a total ban on any use of the hormone on children who fall within the normal range of human variability. Medicine and science ought to be directed to more serious problems.

Nature-Nurture Debate Lives On

*O*ur increased knowledge and manipulative capabilities in genetics carry with them an implicit assumption that hereditary factors, because they are now so accessible, are the major influence on our biological and psychological makeup. The "nature-nurture" question revolves around the age-old debate over the extent to which human intellect, behaviour and emotion are determined by heredity (nature) and how much they are influenced by the environment (nurture).

At the end of the last century and in the first decades of the 1900s, nature was held to be the dominant factor. After the horrors of the Nazi

Holocaust, which was justified by genetic theory, judgement of the scientific community shifted to an overwhelming support for the role of nurture.

The April 13, 1987, issue of *U.S. News and World Report* featured a cover story with the provocative headline "How Heredity Shapes Personality — The Gene Factor." The article reported that the idea of the hereditary basis for much of human behaviour is being revived by studies done with identical twins raised together and apart and by the claims of sociobiologists.

Both groups invoke genetics to explain the complexities of human behaviour and social interactions. They extrapolate from the behaviour of birds, insects, fish and other animals. Once again victims of poverty, racism and imprisonment are being blamed for their own problems: "It's all in their genes."

But the whole nature-nurture question really is a red herring. Of course, we are shaped by the genetic blueprint that we inherit from our parents. Look at the striking physical resemblance of identical twins or the differences between a newborn Asian and Caucasian. To the same extent that there is a genetic component to the range in weight, height and skin colour of a population, then personality traits will also be influenced by heredity; because genes control the structure, function and connections of neurons, behaviour and personality must be affected by heredity. Are we then deprived of free will, driven to act simply because of the gene-dictated neurocircuitry in our brains? Of course not. The evolutionary strategy of our species was the development of a huge and complex brain that has an enormous capacity to learn through observation, trial and error and imagination. It is flexibility in behaviour not instinct that is our species' characteristic.

We like to pinpoint a biological factor as the cause of social problems so that we can design our corrective measures accordingly. This approach has been highly successful in combatting disease. But even here, nurture's role is significant. For example, Harvard geneticist Richard Lewontin points out that all human beings carry the bacilli associated with tuberculosis, yet only a small subset of our population comes down with TB: they are native people, migrant workers, blacks in ghettos and welfare recipients. Poverty as well as the bacterium is a factor in the outbreak of tuberculosis, but we find it easier to deal with the disease itself than to alter the conditions that ensure its occurrence.

Inherent in the claim that something is hereditary is a notion of inevitability, that if it's inherited, nothing can be done. This is a dangerous impression that must be discarded. In studies on temperature-sensitive mutations in fruit flies, my colleagues and I selected such mutations with effects ranging from alterations in patterns of bristles on the body to severe behavioural defects and death. These defective traits appear reliably from generation to generation, *as long as the flies are raised within a specific temperature range*. Outside that range, the flies may be indistinguishable from normal! Other mutations are affected by diet, humidity, drugs, and other factors. We too are the end result of an *interaction* between genes and the environment, and it is a mistake to suggest that when a trait has a hereditary component, there's little we can do.

Every human being on the planet is unique. Even when two people are completely identical genetically, they are distinguishably different in personality and physical appearance. The *range* of the differences between identical twins may not be as great, but experience and environment act on twins' similar genetic makeup to mould unique individuals.

I share genes with people in Japan; our physical resemblance is therefore striking. But the enormous psychological and behavioural gulf that separates me from those people is obvious the moment we open our mouths — environment has been the most important factor shaping our personalities.

The amount to which nature and nurture determine our personalities is a rather trivial scientific question, but the social implications can be staggering, as the victims of the Nazi concentration camps can attest. In Singapore, Prime Minister Lee Kwan Yew has instituted a program of rewarding uneducated people for being sterilized after having one or two babies, while university graduates who have more than two are given bonuses.

As we gain powerful dexterity in manipulating and analyzing the genetic blueprint DNA, we itch to use the technology. In speculating that heredity plays a role in human personality, we begin to assume it is the overriding influence. On May 12, 1987, in Melbourne, Australian reproductive biologist Carl Wood suggested that genetic engineering might be able to eliminate the "killer instinct" in humans, lessen dependence on food, elevate our tolerance of pollution and increase beauty and intelligence. Once again, scientists have become intoxicated with their discoveries and are confusing what they believe with what they know. The nature-nurture debate is alive and well.

There Will Always Be Rushtons

*N*owhere are the hazards of blurring those lines between nature and nurture more real than in issues about socioeconomic disparities between different racial groups. The potential for mischievous extrapolation from scientific data is enormous. This exploded into the public arena in January 1989, when a paper was presented at the annual meeting of the American Association for the Advancement of Science in San Francisco.

The article was written by University of Western Ontario psychologist J. Philippe Rushton, and he has been at the centre of a storm of publicity ever since. Very simply put, Rushton claims that in over fifty categories of physical and behavioural traits, including brain size, sexual activity, family size (he calls them "litters") and criminality, there is a correlation with the time a race evolved. Black people evolved before whites who in turn evolved before orientals and Rushton claims to see the same order when categories such as genitalia and brain size are compared (blacks have biggest genitalia and smallest brains, Orientals have smallest genitals and biggest brains and whites are intermediate). Rushton concludes that the current status of blacks and Orientals at opposite ends of the socioeconomic spectrum is a reflection of heredity, even though the claims that these complex traits are genetically determined has little foundation. Rushton is just the latest in a long line of scientists claiming to prove that nature is a greater cause of social status of a race than nurture.

An academic community supported by research grants and protected from political interference by tenure has a reciprocal responsibility to impose a standard of scientific research and to speak out when scientifically worthless claims are made. In protecting Rushton under the rubric of academic freedom, academics both legitimate and reinforce the racist ideas supported by his claims. This is a terrible abrogation of responsibility because *there is no scientific merit to Rushton's work.*

There are many points of major criticism of Rushton's "data" and assumptions. For example, geneticists have long discarded the notion of *race* as a meaningful biological term. The number of genes regulating the physical differences that make races distinct is very small compared to the amount of variation from person to person within any race. This intraracial variation precludes the term *race* as a useful biological category. Furthermore, Rushton dismisses Inuit and Amerindians who don't fit his scheme as "anomalies"; with such data selection, one can "prove" anything.

But the debate should not get bogged down in the details of the research because the implication is that his work is scientifically valid and it's not. Essentially, Rushton's claims depend on an assumption made by one of his mentors, Berkeley educational psychologist Arthur Jensen, in the comparison of IQ scores of blacks and whites.

Numerous studies reveal that on average black people score significantly lower than whites. Does this difference reflect the environmental disparities between the two races or an inborn genetic difference? Complex indicators of human behaviour such as scores on an IQ test are determined by many genes, and the only way to analyse the effect of genes on the trait is statistically. Geneticists compare the difference in IQ scores between matched pairs of people of varying degrees of genetic relatedness. Thus, if only genes govern IQ scores, both members of a set of identical twins whether raised together or apart should always score the same. They don't, and those results clearly show that the environment influences IQ scores. But there is also a hereditary component — test scores of identical twins are more similar than those of nonidentical twins, who in turn have scores closer together than first cousins, and so on.

When IQ scores of a test population, white or black, are plotted, they fall along the familiar bell-shaped curve with a mean of 100 for whites and lower for blacks. So on the basis of data such as the twin studies, in both populations it is assumed that individuals with similar scores share more genes for IQ test values than people whose scores are farther apart. In other words, nature also plays a role in determining the relative position of one's IQ score. Jensen then concluded that the *difference* between the average scores of whites and blacks must also be influenced by genes. And that's where he exceeded the limits of his study.

Jensen is an educational psychologist, not a geneticist. Shortly after publication of his work in 1969, members of the Genetics Society of America overwhelmingly passed a resolution stating that studies such as his have no genetic validity. The definitive popular article on the subject by Oxford's Sir Walter Bodmer and Stanford's Luca Cavalli-Sforza appeared in *Scientific American* (October 1970). They showed that one cannot extrapolate from observations *within* populations to valid comparisons *between* them. In order to do that, the two populations being compared have to be exposed to the same environmental conditions so that the only variable affecting IQ scores would be race. But racism and bigotry are an overwhelming reality for all black people in North America.

Bodmer and Cavalli-Sforza conclude:

> The question of a possible genetic basis for the race-IQ difference will be almost impossible to answer satisfactorily before the environmental differences between U.S. blacks and whites have been substantially reduced . . . . No good case can be made for such studies on either scientific or practical grounds.

Little has changed in race relations since the definitive statement was made by world-class geneticists almost two decades ago. Why then do people like Rushton persist? One can only conclude that Rushton is either terribly ignorant or deliberately mischievous.

But there will always be Rushtons. Our concern should be over the failure of the academic community to denounce this kind of work as bad science. The fact that the Social Sciences and Humanities Research Council of Canada funds this work and that reputable scientific journals publish the results lends Rushton credibility he does not merit. By defending Rushton's right to pronounce and spread his ideas, the entire academic community is besmirched by supporting shoddy science and stands condemned for a dereliction of its social responsibility.

Our Fragile Democracy

*T*he tacit assumptions about race and behaviour often become evident at the time of a racial "incident." I experienced the results of such assumptions when Japanese Canadians were incarcerated during the Second World War. On April 8, 1968, the Monday after Martin Luther King was assassinated, faculty and students at the University of British Columbia held a memorial service in front of the library.

As one of the speakers, I talked about the lesson King had given us, namely, that even in a democracy you have to fight for freedom and equality. But, I said, Canadians should not feel smug or superior to the Americans, for we too have a sorry history of racism and oppression. I talked about the sad state of Canadian native peoples and went on to remind them about the Japanese-Canadian evacuation and incarceration in British Columbia in 1942. Later, a newspaper editorial chastised me and said that I should not stir up old memories that were best forgotten, lest they revive the antagonism of those who had suffered at the hands of the Japanese.

History provides us with the only real lessons about the fragility of human rights, and today, as the Keegstra trial (for teaching that the Holocaust did not occur and that there is a Jewish conspiracy to control the world) and the Zundel conviction (for spreading anti-Semitic hate literature) remind us, they are so easily distorted or forgotten. Japanese Canadians were insistent that their incarceration, loss of property and forced dispersion should be acknowledged as wrongs and their losses redressed with financial settlements. On April 12, 1985, we had another compelling reason to keep the Japanese-Canadian experience in the public view. On that day, the Ontario Command of the Royal Canadian Legion formally asked the federal government not to make any financial settlement with Japanese Canadians. James Forbes, assistant provincial secretary of the Ontario Command, stated that if Japanese Canadians "are awarded money, then the Japanese government should be pressed to compensate Canadian veterans who were held in concentration camps" (*The Globe and Mail*). But what on earth has the treatment of Canadian citizens by the government of Canada got to do with the treatment of Canadian citizens by an enemy government?

I can't think of a better example of the fragility of democratic ideals. What a tragedy that war veterans, as the one group of people who have put their bodies on the line to preserve our high principles, should be the ones to deny them. When it comes to Asians, Canada has had a tough time living up to its stated civil rights. History informs us of race riots against Asians, racial quotas in university professional schools, denial of the vote to Asians, native born or not, until 1948, restrictive covenants on property — these were just some of the injustices. But we have learned from past experience and now work to eliminate our inequities. The notion that there is only one class of citizen, guaranteed all the full rights of citizenship whether native born or naturalized, must be the goal of this young nation.

The Royal Canadian Legion, at least the Ontario branch, succumbed to the commonly held notion that any rights of citizenship can be abrogated for target groups in times of perceived stress. Yet it is in times of stress that those rights and guarantees matter most. Furthermore, some people appear to believe that guarantees of freedom and equality are for the Caucasian majority. A few years ago, CTV's flagship current affairs show, "W5," demonstrated how seductive this notion is when they reported that foreign students from Asia were taking advantage of the Canadian taxpayer by enrolling in our universities. To illustrate their thesis, they showed pictures of Asian students on campus without regard to whether they were Canadians or not! The assumption seems to have been that if they're Asian they're not Canadians — exactly the sentiment underlying the Ontario Command's submission to the government.

By the reasoning of the Legionnaires, we should press B.C.'s Doukhabors for compensation for the Soviet Union's destruction of the Korean Air flight. Do we blame third-generation Irish Canadians for the atrocities in Belfast? Is every Afro-Canadian responsible for the excesses of Idi Amin? I hope even the Legionnaires would not make such absurd claims.

Then what do we make of their contention that the Japanese-Canadian evacuation is justified on the basis of the terrible treatment of Canadian soldiers in concentration camps in the Pacific theatre? Do they not know that the Japanese were also the enemy of Japanese Canadians? How ironic that Japanese Canadians, denied freedom and equality by their country, nevertheless struggled for the right to join its army. They fought and died for Canada in the Second World War but not for the bigoted nation embodied by the Ontario Command of the Royal Canadian Legion. The Legion's position can only be understood as racism. It assumes that loyalty and commitment are genetically determined rather than acquired socially.

In the interests of democracy in Canada, this bigotry must be exposed at every opportunity. Every time I speak out on the issue of injustice to Japanese Canadians and receive letters or phone calls from bigots who castigate me for daring to raise the issue "after what the Japs did to us," I feel it benefits Canadians to see how fragile our democracy is.

That is why the Japanese-Canadian push for redress had to be pursued vigorously, as an act of commitment to principles for which so many died in battle. It is to the Brian Mulroney government's credit that in October 1988 it formally apologized to Japanese Canadians for their incarceration and announced a $20,000 compensation for every surviving internee.

Racing to Reach the Genome

We have to be on constant guard against the overzealous extrapolation from data in the area of human behaviour. But as our techniques for physically manipulating DNA have blasted ahead, we will be increasingly tempted.

Former U.S. president Ronald Reagan gave the go-ahead to build the Superconducting Super Collider, a $4.5-billion particle accelerator that will allow physicists to probe deep structure of matter. It's an exciting project that will peel away another layer of mystery behind the atom.

And now, biologists have proposed their own megaproject called the Human Genome Project (HGP). The objective of the HGP is to decipher the entire genetic makeup of a human being before the end of the century. It may cost $3 billion to $10 billion to complete.

This initiative is made possible by the astonishing progress in our ability to isolate and manipulate DNA, the molecular blueprint of life containing the information specifying the precise details of how to build the complex structures and carry out the metabolic processes of cells.

DNA is a long chain of four molecular compounds called bases, each having a distinctive shape, which can be likened to letters in an alphabet. They are attached in long strings. Cells "read" three bases at a time. When several hundred or even thousand of these "triplets" are strung together as a single unit, they are called a gene.

When I graduated in 1961 as a fully trained geneticist, the technical achievements now taken for granted were then inconceivable. Back then, we knew that DNA is the genetic material and that humans have twenty-three pairs of chromosomes, but our insights were extremely primitive. In 1961, we couldn't imagine that the genetic code would be deciphered in our lifetimes, yet within four years, all sixty-four triplets were known. We assumed that each chromosome contained several pieces of DNA linked together by non-DNA material. Now we know there's one long molecule of DNA per chromosome.

In 1961, chromosomes could be classified only by lumping them into groups of similar size and shape. Today, each chromosome is readily identified and we can even recognize changes within them. In that year, we thought that genes were simply a series of bases beginning with a "start" triplet (comparable to a capital letter in a sentence) and ending with a "termination" triplet (like a period). Now we know that ahead of and behind the start and stop signals of a gene are long sequences necessary for regulating the time, speed and rate of reading of the gene. We have also discovered that there are long stretches of DNA whose functions we don't understand within a gene itself.

Those are only a few of the enormous insights we've gained since my graduation, and they convey the sense of the great leaps in both our understanding and manipulative abilities. Nothing is more impressive than the tools we have acquired to isolate and determine base sequences of DNA and then to synthesize it, because they are at the heart of the biotechnology revolution. It once took Nobel Prize-winning scientist Walter Gilbert and dozens of his associates years to isolate from a bacterium a specific piece of DNA that was only twenty-four letters long. It was a major achievement when he determined the exact sequence of bases in it. Today, an undergraduate can isolate a piece of DNA and derive a sequence of hundreds of bases in an afternoon lab exercise! In 1977, the complete base sequence of an entire virus DNA was announced — a total of 5,400 bases making up nine genes. It was a major tour de force.

Ten years later, the entire sequence of DNA of a bacterium, *Escherichia coli,* was determined. This represents a description of three million bases! Each achievement has been made possible by improvements in analytical techniques, so that larger fragments can be sequenced faster than ever. The *E. coli* chromosome is about one-tenth the length of the smallest human chromosome, and determining its DNA sequence in turn becomes an achievable target.

The acceleration in DNA-sequencing ability is reminiscent of the way the computer industry has been able to squeeze more and more bits onto silicon chips, at once rendering them more powerful and cheaper. Sequencing of human DNA has been going on in a haphazard way as different scientists have focussed on various genes of interest. At present, perhaps one to two per cent of the DNA in human chromosomes has been sequenced. It will take a co-ordinated effort to catalogue the entire genome within two decades, but the technology to do it exists. All it will take is money and technical commitment.

What is the magnitude of the challenge? The human genome is about three billion bases long — printed out, it would fill two and a half New York phone books. If all the DNA molecules in a single human cell were stretched out end to end, with no magnification, they would measure 2.7 metres! (Incidentally, we are made up of some 100 *trillion* cells, each with 2.7 metres of DNA. That's enough DNA in one person to stretch to the moon and back a million times!)

There will have to be huge data banks to hold and analyse the sequences as they are determined. EMBO, the European Molecular Biology Organization, has such a large computer that is hooked up with Genbank, its U.S. counterpart at Los Alamos National Laboratory in New Mexico. The Soviet Union is interested in the project, and Japan has begun to build a large-scale DNA-sequencing facility under the auspices of government, universities and industry. In 1988, it was possible to sequence up to one thousand bases in a day, but within three years, the Japanese are aiming to do a million a day!

It's an exhilarating project that will yield an enormous amount of information. As Charles Cantor of Columbia University, a leading proponent of the HGP, says, "It will be the Book of Man," a complete description of what makes each of us what we are.

The implications of the project are immense and I hope to live to see it finished. The fact that it appears feasible is a tribute to the stunning technical advances that have been made in the past decades. The HGP will yield a profound understanding of the basis of similarity and variation in individuals. It is estimated that we represent a vast reservoir of variability, differing from each other by about one base out of every thousand. That means, on average, even though we are identical in 99.9 per cent of our sequences, we still differ from each other by three million bases.

When the entire DNA sequence is known, we will come to understand the structure of genes in humans, how they are distributed and perhaps how their activity is controlled. We will gain insights into the mechanism of mutation and evolution and the biology of racial differences. We will be able to infer the structure of 300,000 gene-controlled protein products. The potential for a gold mine of scientific information is exhilarating, and I am as eager as anyone to see the results.

But I am also aware of the socioeconomic and political context within which such work has been carried out in the past. The history of genetics informs us that special-interest groups in society have been quick to seize on new discoveries for their own ends. What genetics does is give us insights into the physical determination of our biological makeup. But often principles governing the inheritance of physical traits such as eye colour or blood type have been inappropriately applied to behaviour and intelligence.

We have long sought clues to what makes individuals what they are, and genes explain our physical makeup very well. And that's what impelled the eugenics movement in the early part of this century.

Eugenics sought the improvement of the human race through selective breeding by encouraging "superior" people to have more children while discouraging "inferior" ones from reproducing. Belief in "biological determinism" — that our lot in life is controlled primarily by our genes — led directly to the enactment of laws in North America for sterilization of people in mental institutions, prohibition of interracial marriages and restrictions on immigration from certain countries whose citizens were deemed inferior.

One-time British Columbia MP A. W. Neill crudely stated this hereditarian doctrine in Parliament in 1941: "Once a Jap, always a Jap." And it was this kind of thinking that resulted in the incarceration of Canadians and Americans of Japanese ancestry during the Second World War. In Nazi Germany, the Holocaust was the extreme result of the eugenics goal of race purification. It was the enthusiastic belief of geneticists that their work was yielding a fundamental explanation of human nature that started this sad chapter in history.

In the past two decades, biological determinism has once again become fashionable. It began in 1969 with Arthur Jensen's paper on IQ differences between blacks and whites. Others soon extended the list of traits purported to have a genetic basis to criminality, poverty, social class and math ability. None of the evidence for these claims is compelling.

What is my point in again raising this uncomfortable history? It is to point out that our current intoxication with the enormous advances in DNA manipulation prevents us from seeing the broader ramifications of what we are doing. The successful sequencing of the entire human genome will yield massive amounts of data that will keep scientists and computers occupied for years. There will likely be little immediate payoff in terms of treatment of hereditary disease, but we will have the illusion that within the DNA sequences resides a comprehensive explanation of all human behaviour and complexity.

With a standard against which individual variation can be compared down to the last base, we will be able to screen for patterns of genetic sequences correlated with different categories of people.

We will search for common elements in DNA profiles of people prone to heart disease, mental breakdown and other ailments. And we could also look at welfare recipients, workers susceptible to disease from chemicals in their workplace, criminals and homosexuals — the list is endless. But an erroneous conclusion is too easily drawn — that people belong to a group *because* they share a certain DNA pattern. It's wrong because the correlation is one based on statistical not causal relationships.

There will be a powerful temptation to attribute poverty, illiteracy, criminality, mental illness and so on to the potential victim's own genes. Would we then bother trying to help them or would we partition them off to avoid "contaminating" future generations? Would we impose preventive detention on people whose DNA sequences foretold public mischief? Who will have access to DNA sequences and how will we avoid misuse of that knowledge? It will be tragic if, in our enthusiasm, we neglect to ask questions that history tells us we must ask if we are to avoid repeating past mistakes.

Final Dance on Racism's Grave

*C*laims for a biological basis for racial traits delight bigots because they appear to give support to the contention that different races should "stay with their own kind." Racism is exacerbated by excessive claims about human behaviour by geneticists. As the media headline sensational stories of racial conflict in our major cities, we should not forget that enormous social changes have been occurring over the past decades.

I walked into my favourite bar a while ago and was hailed by a friend who is a regular there. As I ambled over, I could see that he was involved in an argument with someone I didn't recognize. You could tell their discussion had been heated by the way people in a circle around them were hushed and kept looking over surreptitiously. As I approached, the stranger wheeled on me and demanded, "What do *you* people want?"

Looking back on that moment, I am amazed at the brain's demand for understanding. I hadn't heard a word of their argument and this question was totally incomprehensible, yet immediately my brain began to try to make sense out of it. Faster than you can read this line, my mind had decided that since my friend also worked at the CBC, the issue must have been the recent budget cuts.

"Look," I said, wading in, "there should have been an increase, not a cut."

"No, David," my friend interjected. "That's not what it's about." Instead of helping, that threw me back into confusion, and then my brain began a search down various alleys trying to make sense out of "you people."

My friend used to live in B.C.; could they be discussing the province's economy? Or was it all of us who had gone to university? Perhaps it had to do with everyone who works out nearby at the YMCA. Gradually, my dim wits began to realize that he had lumped my friend and me together because while we are both Canadians by birth, he is of East Indian extraction and I am of Japanese. He meant all of us "coloured" Canadians. What an irony when this guy turned out to have emigrated here from Central Europe when he was a child! Being an articulate academic, I turned on the full heat of my scorn and persuasively told him to "(expletive deleted) off!" and walked away. I showed him not to tangle with someone who has a way with words.

My point in relating this story, in addition to reflecting on how the brain functions to make sense of the world around it, is to share an experience that every member of a visible minority has had in Canada. These days, our media are full of stories of racial incidents. But I think they've got it all wrong. Sure, there is a lot of bigotry here — show me a country where there isn't. But don't let anyone tell you it's worse now than it was.

Each day that I take the bus to work in Toronto, I pass Jesse Ketchum School (it could be any other school in that city) and it fills me with delight to watch youngsters of every colour playing, fighting, hugging — in short, doing what youngsters do but with total unawareness of visible differences. There is in Toronto today a mix of races that would have been impossible to foresee when I was a kid. Back then, most countries of origin of visible minorities were allowed to send one hundred immigrants a year to Canada.

This country is a bold experiment. Modern Canada is grounded in the historic rivalries between those of French and English extraction and the paternalistic oppression of the original inhabitants. There have been a lot of racial horror stories and mistakes and there will be more, but we have learned from past errors, and as we become aware of our often unconscious biases, we do change.

I think Canadians (and I'm one of the guilty ones) are too self-critical — we judge ourselves very harshly. But we shouldn't forget the pluses. My eldest daughter, a fourth-generation Canadian but genetically pure Japanese, never had a moment's concern that her husband is a Caucasian (albeit from Chile), yet in my parents' day, interracial marriage was unthinkable.

My parents, though born in Canada, could not vote until 1948, had to face a quota system in the universities and could not buy property like everyone else. Shortly after the war, my family moved to Leamington, a town in Southern Ontario, where kids would boast that "No black has ever stayed in town past sunset."

My point here is not to decry past injustices but to indicate that what was once considered okay is simply not acceptable today. Yes, there is still a lot of bigotry and we must always fight it when we see it, but I believe for every discriminatory episode, there are dozens of acts of generosity, friendship and assistance that cross racial barriers. They just never make the papers.

I was amazed and filled with pride at the response of Canadians to the Vietnamese Boat People and to the starving Ethiopians. It was generous and genuine. Our record of assistance to the Third World, through agencies like CIDA and IDRC, has been an example for the rest of the world.

And as we grope towards our ideal of equal opportunity and justice for all, we can look to those children in the school ground with hope. I'm reminded of two episodes in parochial Leamington of the 1940s that give me faith in youngsters. One of my great chums went to a club once a week where he had lots of fun playing games. He wanted me to join too, and I was delighted at the prospect. So I tagged along with him, but when I got there, the grown-ups made me sit outside the entire evening. My friend had wanted me to join the Sons of England!

The other experience involves another pal. We were playing together when my dad rode by on a bike. I waved and hollered at him, to the amazement of my friend. "How do you know him?" he asked.

"That's my dad," I replied.

"But he's a Chinaman!" he exclaimed. Yes, kids are colourblind.

TECHNOLOGY'S DOUBLE-EDGED SWORD

Introduction

T he distinction between science — the search for knowledge — and technology — the application of that knowledge — has been clear until the past few decades. In the past, the interval between a discovery and the application of that new finding may have been measured in decades or even centuries. However, in this century, the use of science for warfare — mustard gas, radar, aircraft, nuclear weapons — has meshed the destinies of the military and science. And with increasing hope that scientific R & D will drive the economic engine of society, governments are emphasizing even more the rapid application of new ideas and inventions. So the distinction between science and technology has blurred.

When I graduated from university as a fully licensed geneticist in 1961, we were proud of being "basic" scientists who were carrying out research for its own sake. We were intellectual snobs who looked down our noses at the applied areas of medical genetics or even ecology. But the seeds of change had already been sown with the launch of Sputnik in 1957. The U.S. was catalysed into a catch-up program that culminated in a successful race to the moon. Now we are swept up in another change stimulated by Japan's economic miracle, and once again science is seen as the key to a successful challenge to Japan's lead. But science and technology do not come without problems.

It Always Costs

I have long believed that we have to have greater scientific literacy at all levels of society if we are to have any hope of affecting the way science and technology are impacting on our lives. That's why I went into broadcasting.

But I have only recently realized that my underlying faith in the power of greater awareness is misplaced. First of all, we must understand that there is no such thing as a problem-free technology. However beneficent, technology always has a cost.

Think, for example, of DDT — it killed malaria-carrying mosquitoes in huge numbers and without question saved millions of lives in tropical countries. But geneticists could have predicted that DDT would exert incredible selective pressure for mutations that would confer resistance to DDT in the mosquitoes and that within a few years large numbers would return. They did. But once committed to a chemical approach, we had to turn to other more toxic compounds.

The ecological damage from massive use of chemical sprays has been enormous because DDT is not specific and kills all insects. Furthermore, the compound is ingested by many organisms, so that initially minute quantities become concentrated up the food chain in a process called *biomagnification*. The final result was that DDT ended up in the shell glands of birds, affecting the thickness of egg shells and eventually causing heavy bird mortality.

There are numerous examples of how technological innovations have had detrimental side effects that eventually outweighed their benefits. It has been my assumption that what we needed was some kind of vehicle, like panels of citizens representing a broad range of interests, to do a cost/benefit analysis of all new technologies. The idea was that by carefully weighing the benefits and bad side effects, we could make a more informed decision on whether to allow a new technology to be used. My belief that this would help us avoid future problems was based on faith in our predictive capabilities. Indeed, much of the testing of environmental and health impacts is made on that faith. But we can't rely on such a system.

For one thing, our assessments are always limited. For example, suppose we do an environmental impact assessment of drilling for oil in the high Arctic. The studies, of necessity, are carried out in a limited time within a restricted area. It is simply assumed that scaling up the observed effects of the two drill holes by a factor of one hundred or more gives a reasonable estimate of the impact of major exploration.

Well, there are effects called *synergistic*: several components interact to give new or greater effects than the sum of their individual impact. Also, during an assessment, you can bet industry will be on its best behaviour, so the results will always be on the conservative side.

It is also true that even if a study is made over ten years (which it won't be) we could never anticipate all the fluctuation of conditions in this sensitive area. I've known colleagues who have studied populations of animals or plants over decades and find nice cycles and patterns that are predictable until suddenly completely unexpected fluctuations occur. They get out more publications that way, but we ought, then, to be a lot more humble about how *little* we know.

Finally, we know that major blowouts, spills or accidents are relatively rare. Suppose one happens an average of once every twenty holes. By studying *two* holes and finding no effect, we are not justified in concluding that drilling one hundred holes will also be accident free. It would be just as invalid were an accident to happen in one of the test holes to conclude that half of all drilling sites will have a bad episode. The numbers are statistically meaningless.

Food additives, pesticides and drugs are extensively tested before they are approved for use. But numerous cases inform us that we can't anticipate all the effects. The DDT example is classic — at the time it was used, we didn't even know about biomagnification, let alone its concentration in bird shell glands.

Remember thalidomide or DES? Or consider the oral contraceptive. It had been extensively tested (in Puerto Rico, of course) and shown to be efficacious without detrimental side effects. It was only after millions of healthy, normal women had taken the pill for years that epidemiologists could see negative effects. No amount of pretesting could have anticipated them.

So we come to a terrible conclusion. Technology has enormous benefits. They are undeniable — that's why we're hooked on it. Once technology is in place, it becomes impossible to do without it and we can't go back to doing things the old way. But the pretesting of any new technology is flawed because it provides only a limited assessment of its impact. The tests are restricted in size, scope and time and are based on what we decide a priori might be a possible effect.

But how can we test for something that we don't know will happen? If every technology has a cost, the more powerful the technology, the greater its potential cost. We have to build into our judgements a large leeway for our ignorance and err on the side of extreme caution. And perhaps it's time to realize we don't have to do everything just because we can.

The Illusory Oil Change

*T*he environmental consequences of our technological society with its high consumption and disposable products are forcing us to reassess our lifestyles. Recycling was once the rule, dictated simply by necessity. We have forgotten that our parents conserved routinely and now we are being haunted by the consequences of our profligate ways.

We've all taken our cars in to have the oil changed — perhaps you've even changed the oil yourself. But have you ever wondered what happens to that yucky black stuff that was drained out? That lubricating oil is potentially reusable, but most of it ends up dumped on the ground or in rivers and creeks. It's a classic example of our society's short-sightedness.

We live in a strange world of illusion. The current prices of oil are depressed because of a "glut" on the market, yet every oilperson knows oil is going to run out early in the next century. We have enormous environmental problems, yet we continue to pay no attention to the destructive effects of many of our products that end up polluting. That brings us back to used oil.

There are two kinds of lubricating oil: the stuff we use in our cars, and industrial oils. Of the two hundred million gallons of lubricating oil produced in Canada annually, half is used up in the lubricating process, but the other one hundred million gallons are potentially recoverable. In fact, about thirty-seven million gallons are collected, of which about twenty-two million gallons are re-refined and the remainder is burned or spread on roads. What about the uncollected sixty-three million gallons? Chances are they go into sewers or onto the ground. Without a doubt we end up drinking it in our water and eating it in our vegetables and meat. So while we waste a precious resource by failing to recycle all this used oil, it is also a major contaminant of the environment.

Used oil is also laced with deleterious chemicals that are removed in the re-refining process but that are liberated when the used oil is burned in low-grade furnaces or dumped. Some of those contaminants include lead, zinc, chromium, arsenic, chlorine, bromine, PCBs, polycyclic aromatics and volatile and semivolatile organics — a rather nasty gallery of chemicals.

It costs money to re-refine oil. There has to be a system whereby it can be stored, picked up and transported to re-refining plants. There is little incentive, especially when "virgin" oil — refined crude oil — is so cheap. Twenty years ago in the U.S., there were over two hundred re-refiners. In 1987, only three remained, and they were struggling to survive. At the same time in Canada, there were six, and they were all barely making it.

Part of the problem is psychological — Canadians believe that re-refined oil is lower quality than virgin oil. Yet a study by the National Research Council of Canada showed that re-refined oil is as good as or even better than the refined. But we are reluctant to purchase the re-refined, especially since it is more expensive.

A major part of the problem for re-refiners is political — all the tax incentives and subsidies go to the discovery and exploitation of crude oil. There are no economic incentives for the re-refiners. There should be every encouragement to conserve through recycling and to protect the environment by removing toxic contaminants and preventing the introduction of the oil into the environment. But therein lies the problem. The producer of a product — any product — usually has no obligation to anticipate its *total* cost including eventual disposal, yet that should be built into the initial costing.

The best example of our myopia is the nuclear industry, which built plants long before there was any serious consideration of disposal of radioactive wastes or decommissioning aging plants. Economic or legal incentives to recycle in an environmentally responsible way are needed. Re-refining oil ought to be an apple-pie issue, but the industry is on the ropes.

We have to get over the idea that we can dump liquid waste into sewers and forget about it or that we can slop it onto the ground where it will be absorbed. It is ironic that while the PCB spill near Kenora created a public outcry, *millions* of gallons of used oil — much of it containing PCBs and other toxics — are sprayed onto dirt roads to keep down the dust. More goes into our waters where it is estimated the cost is up to $8 million a year to repair corrosion and replace filtration systems. Millions of gallons of used oil are also burned in furnaces for low-grade heating of greenhouses. The burning temperature is too low to destroy PCBs, which are simply liberated into the air and often end up being absorbed by the plants growing in the greenhouses! Some used oil is actually sprayed on pigs' backs to keep them from getting sunburned in the summer.

Our political and economic systems do not consider the cost of human activity on an ecosystem. Whatever we take from that ecosystem denies it to other life forms, and whatever we dispose into it flows through the various cycles of air, water and soil. For millennia, our numbers were small and our technology simple, so the environment seemed limitless and endlessly self-cleansing. Today we are too numerous and our technology is too powerful for nature to be as forgiving. The economic "costs" of new products simply do not include long-term environmental degradation. Governments don't offer enough rewards to those who conserve resources for future generations or adequately punish those who use up or damage the environment. It doesn't make sense to decide on recycling only if it is economically profitable: we live on a finite planet where all life is ultimately interconnected. It's high time we radically changed our system of economics to take into account our impact on the environment.

So the next time you empty your crankcase, think hard before dumping that stuff down the drain.

Nuclear Menus
(Or, Eating in the Nuclear Age)

*H*istory reveals that every technology, however benefi-cial, has costs. And almost always those costs cannot be anticipated or predicted. For example, one of the great scientific discoveries was penicillin. I vividly remember what a wonder drug it was when I caught pneumonia in the early 1950s. I was terribly sick and my father didn't think I would pull through, so he squandered the family savings to buy me a watch before I died. Then the doctor came and gave me a shot of penicillin, and overnight I was up and around. (I kept the watch.)

But no one anticipated how extensively penicillin would be used and that other "miracle" antibiotics would end up as ingredients in animal feed. Then in the 1960s, scientists discovered a phenomenon called "multiple drug resistance" resulting from a ring of DNA that carries several genes conferring resistance to a number of antibiotics. Inadvertently, the excessive use of antibiotics as prophylaxis and as growth promoters in animal feed had selected multiple resistance that has eventually resulted in the deaths of thousands of people. It is with the knowledge of such unexpected costs that we must assess new proposals like the use of radiation to sterilize food.

How many times have you discovered an old bag of onions or potatoes that are no longer edible because they've sprouted roots or leaves? Ever tried to keep meat from spoiling without refrigeration? Throughout human history, food spoilage has been a constant concern and we've come up with ingenious solutions — smoking, salting, dehydration, bottling, refrigeration, freezing and so on. And now we have the latest, a miracle of twentieth-century technology, irradiation.

Decay is brought on by microorganisms that parasitize dead plant or animal tissue. Storage conditions of dark, humidity and warm temperature may stimulate growth of vegetables. Both of these pro-cesses — decay and sprouting — can be inhibited by massive doses of radiation that kill decay-causing organisms or knock out a plant's regenerative capacity. Food irradiation is being touted as a technological revolution that could feed the world's hungry and, not incidentally, rescue an ailing nuclear industry.

The liberation of energy by splitting atoms had been the crowning achievement of a hitherto esoteric science, and after the war, physicists hoped for an era of cheap, clean, limitless energy by using "atoms for peace." Atomic Energy of Canada Limited (AECL), a crown corporation, has spearheaded the peaceful use of nuclear energy. Canada developed and is the major producer of Cobalt 60 "bombs" for fighting cancer. And our Candu reactor is claimed to be the best in the world as reactors go. But Candu has become an economic black hole: we have poured hundreds of millions of dollars into this technology with no hope of ever recovering the money. A dependable consequence of megaprojects is that once started they develop a life of their own and become difficult to shut off.

Nuclear technology is complex and represents the ultimate practical achievement by science and engineering. But Three Mile Island and Chernobyl put to rest for good the vaunted faith that any technology can be "foolproof." Human beings will always outfool the cleverest failsafe devices. The nuclear industry has been beleagured by public concern about safety, economics and nuclear weapons. So it's not surprising that the nuclear establishment is looking for new justifications and potential income. And that brings us back to food irradiation. AECL and the international nuclear community have been hard at work pressing the case for the economic and health benefits of food irradiation. The Science Council of Canada and leading scientists have strongly supported the claims of safety and the benefits of food irradiation.

Now I have to be honest. People have been pestering me to do an exposé on the hazards of food irradiation for several years. I've been neutral because I have looked at scientific papers on fruit flies fed irradiated food and have not been impressed by data purporting to show that such food is mutagenic. But it's time to get off the fence.

One reason is my inclination to mistrust any grand claims being made by people with a strong vested interest in whatever they're pushing. Few of us are convinced any longer by the tobacco industry's claims about the benefits and safety of its products. Shouldn't we be equally cautious about the zealous sales job being done by people in the nuclear industry? Don't get me wrong — I don't think they are evil or dishonest. But they are *believers* and believers often use every means, including exaggeration, distortion and intimidation to sell the product they have faith in. Ollie North showed us that. We must take the nuclear industry's denial of possible health hazards of irradiated food with skepticism and look to other sources of information (I recommend an excellent article in the March/April 1987 issue of *This Magazine*).

These days we are flooded with drugs, pollutants and additives in our food, air and water. We should know as much as possible about what we ingest. Why then is the food irradiation industry so anxious to hide its activity by deleting any reference to irradiation on food labels? It is insupportable to deny people the right to inform themselves and make up their own minds. If people mistakenly assume irradiated food is radioactive, then it's up to the industry to educate them, not deliberately keep them uninformed.

History teaches that we simply cannot predict the ultimate costs and benefits of new technologies. If any industry should know that, it's the nuclear industry. The long-term consequences of nuclear explosions (fallout, holes in the ozone layer, electromagnetic pulses, nuclear winter) were only discovered long after the weapons had been used. The entire history of technology is full of examples where immediate benefits were obvious, but the costs completely hidden and unpredictable a priori. If we can't anticipate what the effects will be, we don't even know what to look for. I think as long as we get caught up in trying to marshal evidence from animal studies, epidemiology and other areas to show harmful effects, the potential consumer and eventual taxpayer are bound to lose.

The wrong questions are being asked about food irradiation. It is not a debate over whether the practice is safe — we won't know until the technology is widely used and millions of people have eaten the food for years. If history is any guide, there will be unexpected and deleterious effects. Right now a major motivation for food irradiation is to keep an ailing nuclear industry alive. That obscures the fundamental question: do we need a nuclear industry in this day and age? Surely a Royal Commission to reassess the role of nuclear energy in our society is long overdue.

PCBs As Faustian Bargain

Each year, thousands of new chemicals are created and used. Few of them are ever tested for their biological effects, yet they accumulate in the environment and eventually end up in the food chain. No chemical is more controversial than PCBs, and reports about them make me as mad as hell! You can mess around with me, but when it comes to my children, watch out.

The announcements (in fall 1985) from Environment Canada that PCBs could be a serious contaminant in the bodies of our children were the last straw. Our air is now so dirty that we can see it. Our waters are so toxic that the fish in it are poisonous. Farm animals' bodies contain antibiotics, hormones and other man-made chemicals. The litany of horror stories from Bhopal to Seveso to Minamata and the poisons — arsenic, mercury, lead, dioxin, cadmium — have become a sickeningly repetitious part of modern life. And now fruit and vegetables have been found to contain levels of PCBs that could mean that children on a "healthy" diet are exceeding the Acceptable Daily Intake (ADI). The PCB report is not definitive but highly suggestive; we should always err on the side of caution. The calculations are simple and based on the ADI for PCBs of two micrograms per kilogram of body weight per day or sixty micrograms per day for an average-size person. These are some of the discoveries:

A kilogram of vegetables and fruits contains an average of sixty-four micrograms of PCBs.

For a seventy-kilogram person, an average diet of vegetables and fruits gives seventy-three per cent of the ADI.

For a fifty-kilogram person, that works out to 102 per cent of the ADI. For a child, the levels are far in excess of the ADI, and growing rapidly, children are even more susceptible to toxic effects.

Adults, because of our size, have sufficient body mass to dilute the concentration of toxics below the ADI. But the chemicals still accumulate in our tissues, such as the mammary glands of women. And PCBs are only one of many harmful chemicals in our environment.

Today, from conception to birth, a fetus is exposed to a barrage of chemicals never encountered in the past evolution of our species. It may be something as simple as aspirin taken for a headache, a medication for a cough or a decongestant. As well, pregnant women ingest alcohol and nicotine. For a fifty-kilogram adult, an exposure may be within acceptable limits, but for an embryo or fetus weighing a few grams, the effect can be devastating. We use many man-made chemicals at home or work, but we can do something about that.

What can we do when the air, water and food are loaded with toxic chemicals? What alternatives do we have? Now from the moment of birth, a child will be further assaulted by toxic compounds beginning with those in mother's milk! We adults have only been exposed to the buildup in toxins in recent years (even so, it's estimated that eighty to ninety per cent of all cancers are caused by environmental factors), but now children are being hit with these toxins from conception. And rapidly growing cells and tissues of children are much more sensitive to toxic compounds than those of adults. Consider also that it can take twenty to thirty years after exposure for a damaging agent to register an effect. I don't think we can pussyfoot around anymore with weasely statements that "there is no definitive evidence linking PCBs to cancers in humans."

On the basis of animal studies alone, we should assume and act as if PCBs are hazardous until we're convincingly shown otherwise. Right now, we do exactly the opposite: the burden of proof rests with the potential victim. I have no doubt that the health of our children is going to be the price of our intoxication with the power of chemistry.

Agriculture today is on a ruinous course — it has become dependent on the crudest, most short-sighted kind of technology, the application of broad-spectrum agents to kill insect and plant pests. It is blindness to think that this is control and management made possible through an understanding of living things. It is arrogant to assault the very support system of life with our poisons. Many man-made chemicals are not biodegradable — they are brand-new and no life forms have evolved to use them. Hence the compounds and their associated contaminants and breakdown products (which are often toxic as well) continue to accumulate and act over time.

Yet any biologist knows the importance of mutation. In response to the application of chemicals, resistant mutants in plants and insects survive and rapidly accumulate: these are often released from control by natural predators that have been knocked out by the regimen of chemical treatment. So we end up hooked on chemicals as completely as a heroin junkie — we have to have ever more powerful and constantly changing fixes. And when our own crop plants are adversely affected by the pesticides (as they now are), we turn to modern biology's great triumph — genetic engineering — to insert genes to make them even more resistant to stronger sprays. And all the while the chemicals accumulate in the soil and water. Why are we surprised that toxic compounds are not just on the surface but *inside* the fruit and vegetables?

Where are the PCBs coming from? Who is to blame? These are important questions that must not distract us from the overriding point — we've got to kick the habit *now*! It's no longer enough to let industry and health officials demand proof of harm before they will act — that could mean waiting a generation! It's a monstrous blindness if we fail to see where it's leading — to *poisoned* babies! Our current addiction may mean profits for the chemicals industry but damage to the health of our children. Will the crisis of our poisoned environment lead to the ultimate travesty of science when the powerful insights of molecular genetics will be used to "solve" the problem of environmental toxics by genetically modifying *people* to tolerate higher levels? We're already working out the techniques on crops.

What on earth is the matter with us? Not long ago, miners used canaries to monitor the quality of their environment. When the birds died, they paid attention because the air poisoning the canaries was the same air the miners were breathing. For years, the extinction of other species of animals and plants has been warning us that something is wrong. Are we going to wait for our children to manifest multifarious defects before we heed the warning? When we choose the technological road, it takes us on a one-way route beyond constraints imposed by millions of years of evolution into strange territory where we have to keep improvising constantly.

In the 1960s, when student activists were being beaten, gassed, maimed, even killed by authorities, Harvard's Nobel laureate George Wald said to me sadly, "A society that doesn't care for its children is a sick society." And I recently received a poster from the Green Party of Canada with the moving slogan "We don't inherit the earth from our parents, we borrow it from our children." If the threat to the health of our children cannot get us off the poisonous path we're on, what in heaven's name will?

The evacuation of residents of St-Basile-le-Grand, Quebec, in August 1988 after a fire in PCB storage tanks released a cloud of toxic chemicals over the town was yet another reminder to think beyond the immediate questions of health risks, detoxification, culpability and jurisdiction. St-Basile-le-Grand joins a long list of "accidents" involving the unexpected release of toxic products around the world. And it won't be the last. PCBs are a symbol of the Faustian bargain we've struck with technology.

We live in a world of man-made chemicals, many of which have never existed before. Some of them are highly toxic yet are almost indestructible. But we cannot anticipate all the possible long-term "costs" of new compounds because our knowledge about the biological and physical world into which they pass is so incomplete.

Chemicals not obviously corrosive may still represent a considerable long-term hazard. At present, we have no idea what the long-term ecological consequences of chemicals like PCBs and dioxins will be. Only time will reveal the consequences of degradation of the ozone layer by CFCs, the greenhouse effect or toxic chemicals in water on global ecosytems. But we can be sure if other organisms are adversely affected, so will we be.

Thanks to evolution, we have an intricate nervous system that informs us of immediate threats to our bodies through headaches, fever, pain and nausea. But we do not have biological warning devices for radiation or chemicals that attack our genes and cells in more subtle, less direct ways or over long intervals. As we add more and more chemicals and radioactivity into our surroundings, we only find out about their deleterious effects when enough victims provide undeniable evidence.

However clever we are, there will *always* be accidents involving our technologies. The Challenger Space Shuttle and Chernobyl informed us of that. Engineers and scientists are indeed inventive and their creations can be miracles of reliability and efficiency. But even when the mechanical hazards have been reduced to infinitesimally small probabilities, none is ever absolutely safe. That's because human error is the one predictable but uncontrollable element in any new technology. Accidental release of toxic material into our surroundings is inevitable.

We can no longer afford to ignore ecological principles in everything we do. Before any new product or process is allowed on the market, its possible long-term environmental consequences must be seriously considered. Those costs must be an integral part of the market price of any new product. Manufacturers should bear responsibility for environmental and health costs of a product from its inception right through to its disposal. We can no longer afford such short-sightedness as opting for the immediate benefits of a technology without thinking it through to the end.

Think about disposable batteries. As they accumulate in garbage dumps, their nickel and cadmium contents leak into the ground. Yet battery manufacturers have no liability or responsibility for heavy metal contamination. What about the dioxins produced in the process of bleaching paper? For years I have scrupulously avoided using coloured paper products like towels, tissue and toilet paper, only to discover the white products are achieved through the use of highly toxic dioxins and furans. Why must the public pay to clean up lead that has accumulated because leaded gasoline was marketed in the full knowledge of lead's known toxic effects? We cannot afford to profit from foam or plastic products, disposable goods like paper diapers or CFCs in spray cans without first evaluating their social and environmental costs.

St-Basile-le-Grand should also serve as a reminder of how poorly equipped we are to deal with new crises. In small part that is because we don't bother to think things through. It will not be a solution when rain or deliberate washing flushes the PCBs into sewers, the ground or rivers where they can be forgotten. In large part we are enormously ignorant about the consequences of what we are doing. Reassurances by health officials and politicians are usually made to allay public fears when the only honest answer is "We just don't know."

Finally, the sheer weight of human numbers and per capita consumption add up to a massive impact on all the other life forms that share the planet with us. We are expropriating energy, food and space from other organisms, while poisoning their life-support systems with the fruits of our inventiveness. We lack biological controls, drain all life needs from other species and disperse our technological toxins throughout the biosphere.

Self-interest alone should show that our species cannot exist without a thriving community of other organisms. Unless we draw these broad lessons from St-Basile-le-Grand, we will never make any progress in dealing with the environmental crisis exploding around us.

The Lesson of Japan

*T*he astounding recovery of Japan from the ashes of defeat and humiliation of the Second World War has mesmerized politicians and economists everywhere and evoked almost universal admiration. But could it be that our fascination should be replaced by disapproval?

In 1957, when Sputnik was launched, the United States woke up to the fact that the Soviet Union had an educational system that made science its top priority. In response, the Americans revamped their science education and began a massive program to catch up with the Soviet juggernaut. It succeeded beyond expectations.

Now, thirty years later, the Cold War has been replaced by an economic war. And once again North America has awakened to find that another country — this time Japan — has an educational system that appears to be superior to ours and may be the cause of its industrial might.

Japan has galvanized the world with the economic clout it has developed in a few short decades. In response to Japan's brilliant industrial efficiency and the diligence, intelligence and loyalty of its labour force, Western countries are looking to emulate the Oriental powerhouse.

For years, teams of foreign experts have been studying Japanese production and management techniques to find the secrets of their efficiency. Recently, a U.S. group of educators looked at Japanese education and found that the American system compares poorly.

The Japanese story is an impressive one. Where once Japanese products were objects of derision and contempt, they have become symbols of quality. Akio Morita, the chairman of Sony Corporation, has written a fascinating autobiography entitled *Made in Japan*, which provides insights into the Japanese mind. The history of Sony is an amazing story, as it overcame not just the adversities of postwar Japan, but also carved out a niche in the global marketplace against formidable competition.

There is much for us to learn from Morita's book. Most of his opinions and proscriptions seem obvious on reflection. Here are some of them for consideration: North American leaders believe "that if you have a big laboratory with all the latest equipment and good funding it will automatically lead to creativity. It doesn't work that way."

He contrasts U.S. and Japanese ambitions: "Advertisements and commercials in the United States seem to hold up leisure as the most satisfying goal in life, but it is not that way in Japan yet. I really believe there is such a thing as company patriotism and job satisfaction — and that it is as important as money." Speaking of the factors that led to the legendary loyalty and hard work of Japanese employees, Morita says, "It isn't the pay we give that makes the difference, it is the challenge and the recognition that they get on the job."

He explains the special character of Japanese society:

> We Japanese are obsessed with survival. Every day, literally, the earth beneath our feet trembles. We live our daily lives on these volcanic islands with the constant threat not only of a major earthquake, but also of typhoons, tidal waves, savage snowstorms, spring deluges. Our islands provide us with almost no raw material except water, and less than a quarter of our land is livable or arable. Therefore, what we have is precious to us. And that is why we learned to respect nature, to conserve, to miniaturize and to look forward to technology as a means of helping us to survive.

He contrasts this attitude with the North American sense of limitlessness in the expression, "There's plenty more where that came from."

The success of Sony undoubtedly reflects Morita's enormous energy and optimism about technology:

> People are starving not only here but in many other places on Earth, and yet I know that there is the technology available to feed everyone .... Marvellous things are possible if we apply ourselves and we care.

There is no secret ingredient or hidden formula responsible for the success of the best Japanese companies. No theory or plan or government policy will make a business success: that can be done only by people .... Those companies that are most successful in Japan are those that have managed to create a shared sense of fate among all employees, what Americans call labour, management and the shareholders.

Morita's faith in science and technology to solve global problems is boundless:

If we can figure out how to accomplish the task of feeding the world, we may end up with a population problem and a space problem, which might yet lead to another food problem. But I am optimistic enough to believe technology will solve all these problems .... My prediction is that we can enjoy our lives with less energy, less of the old materials, fewer resources, more recycling, and have more of the essentials for a happy and productive life than ever.

These excerpts give us clues to the reasons that the Japanese have succeeded and illustrate their underlying faith in the limitless potential of technology. But are the Japanese an appropriate model for us to admire and try to copy? Certainly from the standpoint of efficiency, diligence and civility, we have much to learn from them. But the people of Japan are totally unlike Canadians. They are homogeneous — racially, culturally, linguistically and historically; Japanese people put a high value on conformity, on fitting in, on belonging.

Canadians, in contrast, are attempting a very different experiment. We treasure, encourage and protect the enormous diversity that reflects a wide variety of ethnic, geographic and cultural roots. We put a high premium on individuality, personal initiative and originality. This creates a lot of problems at times, but the lush variety that results is well worth it.

But there are far more profound reasons to be wary of the Japanese model. I believe that Japan as a country is bereft of a vision of its place within the community of nations and within the global ecosystem, for things said and not said in Morita's book reveal a terrible shallowness of vision and spiritual poverty. The thrust of Japan's economic prowess seems to be aimed at becoming Number One *as an end in itself!* Surely that is an empty goal. That is not an uplifting target for a whole people.

Furthermore, the Japanese can be categorized as a "coloured" race and thus could be a powerful role model and leader of the nonwhite peoples of the world. Yet while leaders from impoverished countries such as India and China have played prominent roles in global politics, Japan seems to be content to focus inward, looking out only to compete with Europe and the United States. The Japanese have a strong sense of racial superiority that blinds them to responsibility to people in the Third World. Canadians, in contrast, have opened their country and pocketbooks to the Third World and should be proud of it.

Morita doesn't mention the Third World until the last few pages of his book, and then he simply advises those countries to copy Japan's example. There is no acknowledgement that global starvation is as much economic and political as it is a technological problem.

However, it is Japan's attitude towards the environment that is most distressing. Morita, like all the uncritical technophiles who trumpet the wonders of the Information Age, believes the benefits of technology are limitless. So there is no consideration of its "costs."

In discussing global starvation, overpopulation and habitat loss, he writes, "I am optimistic enough to believe technology will solve all these problems." Yet technology itself is in large part the *cause* of the problems.

Minamata disease, which is caused by mercury pollution, should have warned the Japanese that costs always accompany technological innovation. And as the only country to have been bombed with an atomic weapon, Japan should be aware of the terrible Faustian bargain technology brings.

Tokyo's Tsukiji, the world's largest fish market, is an astonishing experience. The variety and quantity of sea life sold there are amazing. Most of the seafood comes from elsewhere because the waters around the Japanese islands have been polluted and overfished. Japan's great fishing fleets are like buccaneers of the twentieth-century plying the seven seas to plunder them of enormous quantities of protein. Japanese drift nets form a barrier extending thousands of kilometres down the centre of the Pacific Ocean, intercepting fish, birds and mammals alike. It is sheer "piracy" that threatens the ecological balance of the oceans.

Japan is a resource- and energy-poor country, but with its economic power it has commandeered a disproportionate share of the planet's resources. Thus, for example, Canadian marine regulators, who are under the mistaken belief that they know how to manage the ocean ecosystem, allow herring off the West Coast to be taken in huge quantities at spawning time and sold to the Japanese. Herring is a key species in the support system of larger fish such as salmon, as well as many mammals and birds. But Canadians allow mass "harvesting" of the prespawn fish because their roe is highly prized in Japan. Each female yields a few grams of roe and the rest of the carcasses as well as all males are discarded or rendered for animal feed.

I love herring roe myself, as do native Indians. Because herring are incredibly fecund, within limits we can take some of their eggs without damage. But the vast quantities of herring taken by modern purse seiners (many tonnes per boat) deal a devastating blow to the reproductive capacity of the fish. The ecological value of the live herring populations is immeasurably greater than the revenue from the roe. The Japanese exploit our short-sightedness and avarice by offering money.

But the Japanese are living at the top of the food chain. They have the economic clout to buy what they want or extract from the oceans with impunity. It is neither a wise nor an admirable attitude. As someone whose genes came from Japan three generations ago, I feel sad that such a remarkable people should appear to be so irresponsible, philosophically shallow and bereft of environmental sensitivity. Canadians are leagues ahead of Japan in environmental awareness and a sense of global community.

The Prostitution of Academia

All governments of industrialized countries wish to emulate Japanese success in the high-tech industries, so they are attempting to capitalize on the creative energies of scientists in universities.

Responding to pressure from the government and industry, Canadian universities are encouraging academics to develop ties with the private sector, thereby accelerating the transfer of basic knowledge to industry. The unique role of academic scholars as a group without a vested interest in business or government is thus terribly compromised.

In a glossy advertisement for the University of British Columbia entitled "Engine of Recovery," President David Strangway states on the first page: "Universities are a major source of free enquiry, providing the ideas that can later be exploited by free enterprise. We need both the push of free enquiry and the pull of free enterprise for success in our society." The rest of the brochure is filled with examples of people, primarily scientists, ostensibly solving practical problems in medicine, industry and society.

Across Canada, universities are rushing to become part of the industrial enterprise, as faculty are being encouraged to become entrepreneurs who exploit their discoveries for profit. There have been few objections to or questions raised about this process. I, for one, do not agree with President Strangway's political-economic analysis of the societal role of universities and I have grave concerns about the headlong rush to industrialize the university. Let me explain.

Historically, universities were never meant to be places where people prepared for jobs or where specialists aimed to benefit the "private sector." The university has traditionally been a community of people sharing in the exploration of human thought and creativity. The common assumption since universities became public enterprises has been that if the best minds of our youth are an important natural resource, then universities will maximize their development.

A good university is a place where scholars, dreamers, artists and inventors can exist with no more justification than excelling at what they do and sharing their skills and knowledge. The full range of human thought is encompassed within a university. One consequence is that such knowledge often leads to criticism of government and industry. University scholars can be a pain in the neck to people in power. That's why academics have fought for *tenure* as a means of protection from harassment for their ideas and social critiques. Society needs objective critics if it is to have more than parochial, self-centred goals. Sadly for most Canadian academics, tenure has become a sinecure rather than a privilege and opportunity.

The industrialization of the university is a mistake for many reasons, one of the more trivial being that it will not do what its proponents claim. In rushing to welcome investment from companies to exploit new ideas and discoveries, scientists seem to have forgotten or are unaware that most of our current hotshot ideas will in time prove to be wrong, irrelevant or unimportant. Science is mainly in the business of invalidating the latest concepts. So why the rush to apply them?

But I have much deeper reasons for objecting to the industrialization of the university. The essence of an academic community is the free exchange of ideas, a sharing of knowledge. The formation of private companies within universities and with their faculty runs counter to this spirit. Private companies encourage a destructive kind of competitiveness that can be petty and mean. Secrecy becomes a priority when patenting ideas is a primary goal. And the lure of profit can result in both shoddy science and a narrow focus that ignores broader questions of social responsibility and impact.

My most serious concern is with the vital role of the academic as both critic and source of knowledge for society. Without an axe to grind, the scholar is in a unique position to provide a balanced point of view with data to back him or her up. During the Vietnam War, two of the most visible activists among scientists were MIT's David Baltimore (who later earned a Nobel Prize) and Harvard's Mark Ptashne. They were critical of companies like Dow Chemical and Monsanto for their production of napalm, defoliants and tear gas. Today, both Baltimore and Ptashne have their own biotechnology companies while Dow and Monsanto are heavily involved in biotechnology. Do you think for a minute that Baltimore and Ptashne would be as critical of those industries today? Not on your life.

In the seventies, after the Arab oil embargo, I was involved in a film on the massive deposit of oil in Alberta's Tar Sands. At the time, with oil prices skyrocketing, there was talk of perhaps ten more oil extraction plants as big or bigger than Syncrude. Each would produce at least fifty tons of sulphur dioxide a day. That's a lot of acid rain. So we tried to find a university ecologist in the area who would speak to us on camera about the environmental consequences of such development. We were unsuccessful because no one wanted to jeopardize his grant from the oil companies! Yet it is precisely for that knowledge that society supports such experts in a university.

I don't deny a role for university faculty in the application of new ideas. Our top-notch people are Canada's eyes and ears to the world's research, and good people will have ideas that can eventually be exploited. But the deliberate and urgent push to economic payoff distorts scholarship within the university and subverts its thrust to the will of those who have the money. Profit and destruction are the major reasons for the application of science today, while environmental and social costs are seldom seriously addressed. That's why we need scholars who are detached from those applications.

I remain a faculty member of UBC and because I care so much for the university I am compelled to speak out in criticism. Tenure confers the obligation to do so.

I don't condone but can understand why university scientists, who have been underfunded for so long, are welcoming the Faustian bargain with private industry. But I fail to comprehend why philosophers, historians and sociologists who should know better are acquiescing so easily.

The headlong rush to industrialize the university signals the implicit acceptance of many assumptions that have in the past been questioned by academics themselves. For example, free enterprise, like most economic systems, is based on the unquestioned necessity for steady growth — growth in GNP, consumption and consumer goods.

Steady incremental growth within a given interval is called "exponential growth," and any scientist knows that nothing in the universe grows exponentially indefinitely. Yet economists, business people and politicians assume the explosive increase in income, consumer goods and GNP (and inflation) of the past decades must be maintained to sustain our quality of life. Historians know that this growth is an aberration, a blip that must inevitably stop and reverse itself. But how can the fallacy of maintainable exponential growth be seriously challenged when the university is busy selling the myth that it can help maintain such growth?

Scholars in universities represent tiny islands of thought in society. They are sufficiently detached from the priorities of various interest groups like business, government and the military to point out flaws in our current social truths. But by focussing on issues that are socially relevant or economically profitable, we lose sight of the broader context within which that activity falls; we forget history; we become blind to environmental and social costs of our innovations.

In the U.S., a significant portion of the budgets of universities like MIT, Harvard, Cal Tech and Stanford now comes from private investment. This has split their faculties in debate over whether there should be such close ties with private enterprise. But while those institutions are private, Canada's major universities are all publicly supported. Yet there has been little debate in Canada over the imminent industrialization of academia. The activity and knowledge of our university scientists is paid for by the public and should be available for their benefit, not hidden behind a curtain of classified information, profit priorities or patent secrecy. Academics who accept grants or investments from the military or the pharmaceutical, forestry and computer industries, for example, will be reluctant to jeopardize that support by criticizing those industries when necessary.

There is another consequence of the increased industrialization of our universities that originates in the mentality of scientists themselves. Among scientists there is a hierarchy of position that is directly correlated to grant size and continued research output. A scientist has to keep his "hand in" to maintain status and credibility with his peers. Anyone who decides to look at a wider range of social, environmental or ethical matters, instead of focussing with tunnel vision on specific problems at the cutting edge of research, loses status in the scientific pecking order. Nobel laureates like George Wald of Harvard and Cal Tech's Linus Pauling and Roger Sperry who have become social activists and critics of some areas of science are often referred to disparagingly as "senile," "over the hill" or "out of his area." As university scientists become bound to private enterprise more tightly, their horizons will be restricted even more and they will be far less patient with those who raise social and ethical implications of their work.

Let me be specific by considering one of the hottest areas of applied science — biotechnology — genetic engineering of organisms for commercial purposes. Biotech companies have been sprouting up on campuses like mushrooms. In a number of international meetings held at universities to discuss the future of biotechnology, none has seriously considered the potential misuse or hazards of the technology. Surely an academic community of scholars who maintain an arm's-length relationship with vested interests of society should be expected to raise those questions. Who else will do it?

One of the claims made to encourage greater investment in biotechnology is its potential to "feed the world's hungry." It is a self-serving, shallow justification. Starvation on this planet is a consequence more of political and technological factors than a shortage of food. Even if it weren't, the exponential growth of our species' numbers, which has already doubled the global population twice in the past century, will far outstrip any increase in food production brought about by biotechnology. Scientists anxious to justify their research for more support will resist such objections.

Canadians should be wary of the uncritical push to increase the links between unversity academics and private industry because there are unacceptable "costs."

A Bid for Scientific Excellence

*I*f we look at those areas where there has been a successful relationship between university personnel and the private sectors, it is most obvious in Silicon Valley in California and Route 128 in Massachusetts. These areas of high-tech enterprise are adjacent to world-class universities like Stanford and Berkeley in California and Harvard and MIT in Boston. When there is a community of scholars, important and useful ideas can be expected to eventually spin off into the private sector. But how do we build such a community?

In a world dominated by the U.S., Japan and Europe, we should remember that Canadians have supported a scientific community at starvation levels for decades and we continue to underfund our scientists. At present our research output is about three to four per cent of all of the work done in the world. There are many reasons why most of the money committed to R & D in the provinces and federally will not produce the economic benefits so optimistically proclaimed. I'll focus on one.

Across the country, politicians are allocating research money with a very big string attached. We no longer adhere to the belief that science is a cultural activity a society supports to affirm that it is a civilized one. Research is funded to make discoveries and inventions that will be useful or economically beneficial.

The effect of pressure to apply research knowledge has overwhelmed the scientific community and now dominates both the criteria for granting funds and the way scientists justify their work. Everyone is looking for a practical angle that will legitimize experiments. This perpetuates a mistaken notion of how science works, but that's the subject of another essay. I believe it is extremely short-sighted to carry out research with a view to its potential use because a severe constraint is imposed on the kinds of problems attacked and experiments done. In the current funding climate, a lot of projects with enormous long-term value will not be supported or will be abandoned. Let me give you two examples of past research that would not have received much support had today's standards been applied at the time.

The first involves work by the Swiss microbiologist Werner Arber. He spent years studying a puzzling phenomenon whereby the properties of viruses were modified after they infected certain kinds of bacteria. He discovered that in the bacteria, the virus's genetic material is chemically altered in a way that does not change its genetic content but does make it resistant to certain enzymes that would normally break it down. It was an esoteric study that seemed to elucidate a special quirk of microorganisms that wasn't very relevant to the central problems of genetics.

We students in genetics studied Arber's work because of the elegance of his research. The puzzling phenomenon revealed the existence of a class of proteins that recognized particular DNA sequences and cut the molecules at that point. Back in the early sixties no one anticipated that this would lead to the discovery of tools that make genetic engineering possible, yet that's what happened. Those proteins, called "restriction enzymes," have become the critical means of chopping DNA into smaller fragments. Arber shared the Nobel Prize for that work, but had he been required to justify his research when he started, even he could not have foreseen its eventual utility.

As a graduate student, I remember sweating over the research papers of a highly acclaimed scientist who worked with corn. Barbara McClintock was a famous American geneticist whose studies into heredity in corn were at once elegant and complex. She discovered a puzzling phenomenon in these plants: under certain genetic conditions in certain parts of the plant, genes seemed to *jump* from place to place on chromosomes. This made no sense and ran counter to accepted genetic orthodoxy that each gene occupies a specific spot on a chromosome and is extremely stable. We paid attention to McClintock's work back in the fifties because her experiments were impeccable and she had a world-class track record. But none of us believed that the existence of jumping genes was more than an aberration of the special genetic properties of the corn plant.

Only after more than twenty years passed did scientists working with other organisms discover the phenomenon of such "transposable" genes. With the expanding development of techniques for analyzing DNA, not only did McClintock's observations became acceptable, but also her models are the basis for the fancy tricks now performed by genetic engineers. She was awarded a Nobel Prize for this work, but if she'd had to contend with today's demand for immediate usefulness, I doubt that a funding agency would have taken a chance on her work when she started.

There are many lessons for Canadians in the stories of Arber and McClintock. Perhaps the most important is that outstanding people do interesting and top-notch work. History shows us that we should invest our money and faith in the best people regardless of what area they choose to pursue. We should judge the quality of the scientist, not the relevance or practicality of the proposed research.

Canadians do a tiny fraction of the research done in the world. We are foolish to try to jump in on current "hot" areas of research like biotechnology, computers and superconductivity for immediate short-term applications. We don't have a scientific community that is large and flexible enough or funding levels to get us competitive internationally in a short interval. We have to take time to build a world-class community of scientists irrespective of area, fund them at a level that makes them competitive anywhere and have faith that they will make important discoveries.

Another lesson: the path that scientific discovery follows is unpredictable. The essence of discovery is its constant surprise — that's why most of us go into science, to find out the unexpected. It's impossible to predict what research area will eventually prove useful. Do we seriously believe that if we are fully funded for our grant proposal, we will end up with the solution predicted? Of course not — that's not how science works. Scientists should speak out against the myth that money committed to research leads to discoveries and applications in a linear, predictable way. And if politicians are serious about using the taxpayers' money wisely for social benefit, they'd better educate themselves about the nature of science and discovery so they can spend the money wisely. There are other reasons why the overwhelming bulk of that investment will waste taxpayers' money.

Although there is a terrible need for more money to support science in this country, the *way* the money is being spent reveals a total lack of understanding of how to build a long-lasting, effective research community. Scientists must be assured of steady funding that allows them to be competitive with their peers in any other country. And they must know that they are valued for their excellence *regardless* of their area of interest. Assured of long-term funding for good work, scientists can explore wherever their experiments lead them, attract scholars from around the world, attend international meetings to hear the latest research and unexpectedly spin off ideas that will prove useful to society. Bureaucratically dictated priorities do not lead to useful discoveries.

Canadian governments have starved the scientific community for decades. It's a constant struggle to acquire the funds and equipment to compete internationally, and ironically, a few manage to do so with American grants! When I returned to Canada in 1962, my first Canadian grant was for $3,600 at a time when my American peers were starting off with $30,000 to $40,000! What made it possible to get my research off the ground was a five-year grant that I received from the U.S.!

Once our lab had established a reputation and we were receiving relatively large grants, I was told by a visiting grant committee that we deserved to get every cent we had applied for, but that if we got it, our grant would wipe out ten smaller ones. So we didn't get what we had asked for. Top scientists must be able to get what they need without draining the rest of the scientific community.

And that's only the beginning. If Canadians are to compete in the tough arena of the international marketplace, we *have to believe in ourselves*. In the debate over Free Trade in 1988, Canadian industrialists and entrepreneurs indicated their eagerness to get at the huge American market. To sell what? Look at the goods in our homes, garages or workplaces — chances are only a tiny fraction were manufactured in Canada. Our main attraction for Americans remains our vast resources and space, not our manufactured goods.

For years I wrote letters to Mr. Trudeau to convince him that our greatest need is faith in ourselves. Books are full of examples of world-class ideas created by Canadians who had to go elsewhere to find the backing to bring them to market. (One of the earliest books on the subject is *Ideas in Exile*.) Without self-confidence, when we open the gates to American entrepreneurs, we'll ape them rather than setting our own course. But we don't build that self-confidence by simply proclaiming our intent to compete internationally and beating our breasts about how good we are.

Generations of youngsters have to grow up with examples of people in Canada who have made outstanding contributions. Recently several books on Canadian "heroes and heroines" have been slapped together for school kids, but they lack any critical criteria of excellence and are more often compiled with an eye to a proper balance of sexes, ethnic groups and geographic distribution. They are simply not credible. I wonder how many high school science *teachers* could name a dozen Canadian scientists of the calibre of John Polanyi, Tuzo Wilson or Sir Frederick Banting. We need those role models for our youngsters so they can aspire to similar achievements and feel that they have a fighting chance to do it in their own country.

In 1972, a grade 10 science teacher came to me at the University of British Columbia and told me he had a student who was too bright and keen for high school. The teacher asked me if I could help. The boy was Gordon Wong, a second-generation Chinese Canadian. He *was* very bright and I offered him a summer job, which he accepted and filled very well. The next fall, he moved to a high school close to campus so he could continue to carry out experiments in my lab after school and on weekends.

In his last year in high school, Gordon sat in on my third-year university genetics course, took the exams and scored one hundred per cent on one that over half the class flunked. He ended up with an honours degree in physics and genetics and won a Rhodes scholarship to Oxford University, where he earned his Ph.D. in molecular genetics. Gordon went on to do postdoctoral work at Harvard with a leading molecular biologist, then took a position with the Genetics Institute (GI), a top biotechnology company in Boston. I am very proud of Gordon, but it grieves me that he will probably not return to Canada because he simply cannot find an atmosphere in Canada that duplicates the excitement and excellence that he has at the GI. *That's* what's missing here and that's the challenge for politicians if they mean what they say. There are too many Gordon Wongs who represent the loss of some of our best minds and potential role models for our up-and-coming scientists.

Long ago, my father taught me that we are what we *do*, not what we *say*. It's a standard of judgement we should apply much more, especially in the world of politics. What are politicians doing for science?

Prime Minister Brian Mulroney visited the University of Waterloo on March 4, 1987, and made some stirring remarks about the central place of science in the Canadian economy. He announced that his government is going to ensure that we will be able to compete for part of the global high-tech market. He paid lip service to the fact that science is the main source of technological innovation today, but he has given no indication that he means what he says or knows how to help it.

Politicians trot out science when it is convenient, so the scientific community has been yanked up and down, held hostage to the political climate of the moment. The current government is no different from its predecessor in its support of science, but Mr. Mulroney's party ran on the promise of raising research and development support to a level comparable to that in Japan, the United States and West Germany. In those countries, R & D takes more than 2.5 per cent of the gross national product, while Canada's support is about 1.5 per cent.

Since achieving office, the Tory government has made reduction of the deficit its main goal, and science has had to absorb *cuts*! Thus, spending has been severely cut back at many government labs while areas designated as priorities (such as space research) have had an infusion of support. That is the government's right, of course, but it can't have it both ways — cutting back on an already undersupported group while demanding that the group be globally competitive.

Since Canadian scientists do less than four per cent of all of the research around the world, the chance that one of them will make a big breakthrough is correspondingly low. So then, do we agree with Bud Drury, who when he was Trudeau's minister of state for Science and Technology, told me science is supported only as a frill? Most definitely not. The price of a front-row seat at the "theatre" of discovery is having a community of world-class scholars.

Our top scientists are Canada's eyes and ears to the vast body of research going on around the world. Without them, we are blind and deaf and doomed to buy other people's inventions because we can't capitalize on ideas quickly enough. It is silly to think that Canada will suddenly carve out a place in the international market by identifying "strategic" areas and infusing them with a bit more money. If we have a small group of outstanding researchers, they will keep us plugged in to what's happening everywhere else.

Politicians like to set up institutes, construct buildings and buy equipment. When scientists were begging for more money for operating grants, then-prime minister Pierre Trudeau set aside $51 million in capital funds to construct the Biotechnology Research Institute in Montreal. That's how he thought we would get into a hot area. In British Columbia, former science minister Pat McGeer established Discovery Parks adjacent to four B.C. institutions of higher learning, as if they would become counterparts to Silicon Valley. Trudeau and McGeer had it all wrong.

It is *people* who innovate. We must support imaginative, original, bright people. Building a world-class community of scientists is not like setting up a factory to make radios or shoes. Scientists must have a climate in which they feel valued, supported and respected. Over time, they will attract other scholars by the quality of their work. A centre of excellence will evolve out of which ideas and applications will flow.

Unfortunately, politics demands payoffs on investments before another election, a time frame too short for good science to take hold and flourish. It's far better to establish an institute or build new laboratories as tangible proof of what politicians are doing. And that is why I'm so skeptical when I hear yet another inspiring speech. It's all just *words*.

SCIENCE AND THE MILITARY: AN UNHOLY ALLIANCE

Introduction

Scientists are reluctant to acknowledge that the major uses to which their disciplines are put are *military*. Over half of all scientists and engineers in the world work directly for or with research grants from departments of defence. While this application of new insights for destructive purposes is not new, the scale and scope of modern military establishments is without precedent.

Today, the total global budget for military expenditures amounts to the staggering sum of $1 trillion (U.S.) annually. From that budget, monies for research from the Defense Department in the United States, for example, are twice as great as the combined amount spent for transportation, medicine, basic science, communication and agriculture. And whoever pays the piper calls the tune in terms of priorities. It is a grotesque perversion of the high ideals of the scientific community for its ideas and discoveries to be used for destruction and death.

Defensive Biology

*T*hroughout history, human inventiveness has given us knowledge with which to affect our environment and ourselves. All human knowledge has a double-edged potential to be used for human benefit or detriment. We've used the spear, the knife, the bow and arrow to feed or to kill ourselves.

The bombing of Hiroshima signalled a major change in the course of human affairs. Physicists had come to understand that matter and energy are interchangeable, so there are vast stores of potential energy within atoms. As a result of that insight in basic science, the best scientists of the day conceived, sold and built the most awesome weapon ever imagined — the atomic bomb. And ever since, science and the military have been inextricably linked. We cannot imagine modern warfare without high technology or science without the massive support by defence funding.

Science played an important military role even earlier than the Second World War. During the First World War, chemists discovered the terrible destructive power of gases such as chlorine, phosgene and mustard. The compounds choked and burned people who came into contact with them and were used to kill more than a million people. Death by gas was seen to be so terrible that in 1925, forty nations signed a Geneva protocol banning the use of chemical weapons.

In 1936, a German chemist discovered nerve gases that enter the skin and kill by blocking the transmission of nerve impulses. Although never used, a quarter of a million tonnes of nerve gas was stockpiled by Germany. It took the United States almost fifty years to sign the Geneva protocol. But the U.S. government made "binary weapons," which effectively get around the protocol. In a binary chemical weapon, two chemicals that alone are relatively harmless are stored in separate compartments within a shell. When fired, the spinning shell forces the chemicals to combine and react to form a deadly compound, in this case, nerve gas.

Now, it is the biologists' turn. In the past, people practised the antecedents of biological warfare. They poisoned well water, destroyed crops and livestock and hurled diseased corpses and infected blankets over city walls. But modern microbiology has enabled the culturing of some of nature's most deadly diseases. Biological warfare labs have looked at a wide range of viruses, bacteria, fungi and other organisms that infect people, plant crops and livestock. In 1944, British scientists infected Gruinard Island off Scotland with the deadly spore that causes anthrax. The island remained contaminated until 1986 when authorities finally began decontamination.

In 1972, then-president Richard Nixon signed the Biological Weapons Convention, under which all research on biological weapons was stopped and all cultures were destroyed. Today, there is renewed concern among scientists about the hazards of a race for biological weapons. That is because the 1972 convention was signed when the new recombinant DNA technology was in its infancy. Today, genetic techniques are incredibly sophisticated and make possible the introduction of genes controlling deadly toxins or viruses into bacteria.

During the Reagan administration, there was an ominous increase in defence spending on biological weapons — from $14.9 million at the beginning of Mr. Reagan's first term in 1981 to $73.2 million by 1987. Inspection of a list of the recipients of defence grants reveals some of the leading scientists in biotechnology, including the distinguished Nobel laureate Gobind Khorana of the Massachusetts Institute of Technology. It is clear that the military, by financing top scientists, is keeping an eye on the latest ideas and techniques in molecular biology. Scientists in private companies and universities in twenty-one states are receiving military contracts for biological weapons. They are studying the deadly dengue fever virus, Rift Valley fever virus and other organisms that cause Japanese encephalitis, anthrax, Rocky Mountain spotted fever, leishmania (a parasitic disease) and dysentery. There are also studies on snake venom genes and botulinus toxin.

The military argues that the research is defensive and concerned primarily with the development of vaccines and diagnostic techniques. It is a spurious argument. In carrying out the research, scientists must learn how to culture the organisms and determine their life cycles. Indeed, there have to be offensive weapons to test the defences.

The terrible lesson we learn from history is that there is nothing so horrifying that the military will not pursue it. Often such pursuit is rationalized as necessary because of the enemy's inhumanity.

It is time the scientific profession established a code of ethics for the conduct of its members. Individual scientists have taken stands on work with nuclear weapons, genetic engineering and the Strategic Defense Initiative, but there should be a code governing us all. The greatest deterrent to the development of terrible weapons is public opinion *before* they are built. *Secrecy* hides this dark side of human imagination, and every scientist should do her or his best to make known the potential hazards as well as the benefits of her or his work.

A Technology Out of Control

The nuclear genie has not gone away. Once let out of the bottle, nuclear weapons have become ever more powerful and their delivery mechanisms more sophisticated. A visitor from outer space who was not aware of the psychology of fear, ignorance and aggression that underlies war would likely conclude from observing our weapons race that we are an insane species. There is nothing rational about a defensive program based on the notion that a country with the capacity to destroy its enemy thirty times over has an "advantage" over its opponent whose destructive power can "only" kill us twenty times.

Just as impressive as the killing power is the system for delivering bombs — fast, accurate and deadly — and beyond human capabilities for control. Like everyone else, I've been thinking a lot about nuclear weapons lately and I would like to make three points that I think have yet to be considered.

1) Many prominent scientists are at the forefront of the movement to halt the arms race, but I have yet to hear one of them acknowledge that scientists may be especially culpable for the insane spiral of armaments.

Philip Morrison, who actually armed the first American atomic bomb and was one of the first Americans to see Hiroshima after the A-bomb blast, is a passionate opponent of nuclear armaments, yet he bristled in disagreement when I suggested that scientists had a special responsibility. But I stick by that claim. I don't believe that military strategists come up with the ideas for neutron bombs, binary nerve gases, new biological weapons, particle beam space platforms and laser devices. They come out of the imaginative minds of scientists and are created by engineers and technologists, people who are often our colleagues and our students.

There is intense pressure on science students in university to focus only on science, which precludes any time for other courses such as philosophy, literature, religion or history. The result is that few scientists and engineers today are aware of the history of their professions, a history that might put their current activity into a different social perspective.

The morality and ethics of science and technology and the social responsibilities that accompany professional careers are not matters of discussion during our training. The implications of the tight link between research and applications for military and industrial use are simply not considered by ambitious graduate students. Scientists have a lot of thinking to do about their own peers when it comes to the nuclear issue, and perhaps such thinking could eventually evolve a code of ethics for scientists.

2) My second point concerns the nature of the nuclear weapons debate and the part that personal and emotional elements play in it. If you have ever talked to someone whose position is different from your own on abortion, religion or politics, you know that the factors influencing such positions are seldom based on reason. This is evident when two great scientists, Linus Pauling and Edward Teller, use the same data to reach opposite conclusions on nuclear arms.

In arguments, one person is often heard to exclaim, "You're being too emotional," or "You're taking this far too personally," as if emotions or personal involvement are not relevant to the debate. If issues were simply problems like mathematical puzzles, they would be readily solved with an unequivocal single answer. But human problems are incredibly more complicated by the emotional, nonrational, culturally shaped reactions that tend to overwhelm the analytical functions of the brain.

I don't think we pay enough attention to this reality. In all the talk about arms limitation and defence strategies, people are depicted as rational creatures. But the mistrust and suspicion that lingers between the two superpowers must be seen in the historical and personal backgrounds of the leaders involved. In the end, the biases, the limitations in experience and knowledge of individual human beings will be overwhelming factors determining the fate of nuclear weapons.

3) Finally, the technology to deliver nuclear weapons is now so accurate and fast that it is literally out of human control. All military planning with respect to nuclear weapons is based on the assumption that both offensive and defensive strategies are rationally thought out and acted upon. I submit that it is impossible for people to respond as planned. Consider this:

The deployment of the Pershing II and SS-20 missiles in central Europe now brings targets within a ten-minute range. To cope with that stark reality, about three-quarters of the 4,000-odd satellites in place are there to spy for the military. In principle, the satellites can detect a missile launch seconds after it leaves its silo and can relay the information to military headquarters at the speed of light. Both the Soviet Union and the United States have invested billions of dollars to build "supercomputers" for the military. Yet over the past decade, NATO's computers have mistakenly identified flying objects as missiles more than 150 times!

Of course, it can be argued that the system must work because there hasn't been a nuclear launch, but that is hardly reassuring in the knowledge that we can only have one mistake. The supercomputers are supposed to assess the nature of the attack instantly — the payloads, trajectories, targets and probable damage. Yet we have to assume that several minutes of the ten-minute response window will be used by human brains attempting to verify the computer's analysis, to assimilate the implications and to take action.

If a deliberate attack is ever launched, I would expect it to be started at an "inconvenient" time by the enemy, say at 3:00 a.m. Christmas morning. When the commander-in-chief is finally roused with the news, can we assume that he or she will be (a) instantly awake; (b) able to ignore concerns for personal safety, family possessions, the world; (c) capable of comprehending the information, the options, and the consequences; and (d) lucid and confident enough to make a decision on which the fate of the world may rest?

This simple scenario shows the absurdity of all the military planning — the technology of nuclear weapons is out of control because human beings simply cannot function quickly enough under the incredible pressures, time constraints and implications.

Since the military thinkers must realize this, we have to assume that the supercomputer is programmed to analyse the incoming data and to determine the options and their consequences from moment to moment. In an attack situation, very quickly in a ten-minute interval the machine will indicate that there is only one option left and that it must be acted on immediately or it will be too late. Since it is highly unlikely that a person could make such an awesome decision, then the programmers must put the final decision in the control of the computer. Then the technology is literally out of human control.

The Ultimate Technological Fix

*U*ndaunted by our failure to design a foolproof technology, our technocrats remain filled with faith in their ability to correct all our problems with still more technology. And in the military, that means the Strategic Defense Initiative.

In 1985, SDI was frequently in the media as our government attempted to formulate a policy to reconcile the two highly polarized positions on Canada's involvement.

Polls indicated that the majority of Canadians did not want to become caught up in U.S. President Ronald Reagan's Star Wars dream. But often politicians find it difficult to weigh moral and ethical considerations about escalation of the arms race against more concrete suggestions of potential jobs, promises of high-tech spinoffs in the civilian sector and concerns about Soviet military advantages.

Although history shows that the more complex a technology is, the greater the number of glitches and error potential, technological optimism is rampant today. But it is hard to ignore the arguments of a scientist who does defence research for the U.S. military and nevertheless argues that SDI won't work.

Such a person is David Parnas, formerly Lansdowne Professor in Computer Sciences at the University of Victoria and now professor at Queen's University in Kingston, Ontario. Parnas is a distinguished scientist who turned down a job at Harvard to come to Canada and was appointed on June 5, 1985, to a U.S. Defense Department panel on "Computing in Support of Battle Management."

On June 28, 1985, he resigned because he had concluded that the goals of SDI could not be achieved. He wrote:

> In March, 1983, the President asked us, as members of the scientific community, to provide the means of rendering nuclear weapons impotent and obsolete. I believe that it is our duty, as scientists and engineers, to reply that we have no technological magic that will accomplish that. The short-term applied research and focused development that SDI is now funding is not going to solve the problem; the president and the public should know that.

In a series of eight short papers, he set out in simple form the reasons underlying his decision.

Now before I go into this, let me point out that Parnas has not been a past critic of SDI or defence research. He has been involved in research for the military for over twenty years. Parnas has an appointment (at least when this was written in August 1985) with the Naval Research Laboratory in Washington, D.C.

Nor is he averse to receiving some of the vast sums of money being spent in this area. His conclusions are based on his knowledge of the computer software programs that SDI will depend on.

As he says in his letter of resignation:

> For many, the project offers a source of funding; funding that will enrich some personally, while offering others new and generous support for their personal research projects. During the first sittings of our panel, I could see the dollar figures dazzling everyone involved. Almost everyone that I know within the military industrial complex sees in SDI a new "pot of gold" just waiting to be tapped.

SDI is a project that is built on faith — tremendous faith in the power and reliability of technology. In a sense, we made that leap of faith when the commitment was made to harness the brainpower of physicists to develop the atomic bomb. And it paid off.

Today, we are every bit as dazzled by the achievements of microelectronics, so much so that nothing seems impossible. The conceit behind SDI is the belief that the awesome array of weapons and their delivery systems can be monitored by thousands of spy satellites, that supercomputers will be able to assess the input from those surveillance systems and that now a technology to intercept the incoming missiles in space can be invented.

Let's ignore the many persuasive arguments that particle beam or laser weapons of sufficient energy cannot be built in space, that even if they were, they will be easily deflected or eluded, or (as even the most optimistic proponents concede) that they will never be 100 per cent effective. Let's concentrate instead on the component that will make or break SDI — the software of computer systems.

The computers must handle an enormous amount of information pouring into them, accurately assess it in seconds, direct an immensely complicated interception technology to hit targets travelling through space at thousands of kilometres an hour, while distinguishing between a bewildering array of decoys and evasion tactics. It is estimated that the program for SDI will require nineteen to thirty-five million coded lines (my very sophisticated word processor has about 5,000)!

I don't pretend to understand the mathematical basis of software programs. But I do know that software is designed by human beings who must attempt to anticipate all the requirements that will be imposed at the time of use. Even the simplest programs, if used extensively enough, turn out to have inadequacies or errors, while the more complex the program, the greater the likelihood of "bugs." The only way to get rid of bugs is exhaustive testing, but Parnas points out: "Even after such testing we have incidents such as occurred on a space shuttle flight several years ago. The wrong combination of sequences occurred and prevented the flight from starting. The military software that we depend on every day is not likely to be correct." This is terrifying when you consider that those computer programs are already controlling the satellites and missiles that are now deployed.

But why will the SDI software not be trustworthy? Here's Parnas:

The system will be required to identify and track targets whose ballistic characteristics cannot be known with certainty before the moment of battle. It must distinguish these targets from decoys whose characteristics are also unknown. The computing will be done by a network of computers connected to sensors, weapons and each other, by channels whose behaviour, at the time the system is invoked, cannot be predicted — it will be impossible to test the system under realistic conditions prior to its actual use. The service period of the system will be so short that there will be little possibility of human intervention and no possibility of debugging and modification of the program during that period of service.

Nevertheless, software systems will be designed by people who believe in SDI or at least are willing to give it a try. Here Parnas warns:

> Fire control software cannot be written without making assumptions about the characteristics of enemy weapons and targets. No large-scale software system has ever been installed without extensive testing under realistic conditions. Even with these tests, bugs can and do show up in battle conditions. It is not unusual for software modifications to be made in the field. Such opportunities will not be available in the thirty-minute war — difficulties in building software increase with the size of the system, with the number of independently modifiable subsystems, and the number of interfaces.

In order to overcome these problems, the military mentality will strive to develop further technological aids — programs to simulate actual battle conditions, perhaps even a small-scale controlled attack, and of course, the big promise is artificial intelligence — a machine that is so smart that it will debug the system for us. It's sheer madness.

Can we not see that this is a dead-end reliance on a complex and imperfect technology that leads us into more elaborately contrived ways to overcome the limitations? Surely it would be better to use a tiny fraction of the money spent on weapons to develop novel ways to monitor adherence to weapons treaties. We should seriously negotiate with the Soviets to end the wasteful, draining enterprise of weapons development so that resources and brainpower can be invested in more immediate and rewarding civilian problems. It's not an unrealistic dream to talk of the need for an immediate weapons freeze and negotiated de-escalation — the technological paraphernalia are now so dangerous that we have to.

Parnas presented a devastating analysis that External Affairs Minister Joe Clark and his committee simply cannot ignore. As a scientist, I am equally disturbed by Parnas's description of the facility with which our colleagues — scientists and engineers — are ready to ignore conventional standards of scientific judgement. When he submitted his letter of resignation, Parnas was greeted with two types of reactions: the first was the admission that the project was impossible, but the rationale was that the money was going to be spent anyway, so why not get in on it; the other was a mocking and macho version of "What's the matter, Parnas? Don't you have the guts to take up the challenge?"

But the rot at the heart of the project is money. Parnas wrote in the paper accompanying his letter of resignation from the panel:

I have been astounded at the amount of money that has been wasted on ineffective research projects. In my first contact with the U.S. Navy, I watched millions of dollars spent on a wild computer design that had absolutely no technical merit. In computer software, the DoD [Department of Defense] contracting and funding scheme is remarkably ineffective because the bureaucrats who run it do not understand what they are buying.

Parnas told me that much of what passes as defence research is "scientific fraud" and says that if Reagan were to have suggested that we should build a weapon that travels faster than the speed of light so it would hit the enemy before they can see it, scientists and engineers would have lined up with proposals to work on it!

As Canadians, we should be very proud to have someone of the integrity and courage of David Parnas, and we should also be smart enough to take his advice and steer clear of SDI.

Ethnic Weapons

*F*ormer U.S. president Ronald Reagan signalled his interest in using the latest ideas in molecular biology for war by increasing the budget for biological weapons research fivefold during his tenure. Once again, it served as a warning of the vast potential to apply scientific knowledge for human benefit or detriment and the willingness of political and military leaders to consider even the most repugnant devices.

Horrifying images of Kurdish victims of poison gas and the controversy over Libyan production of chemical weapons reminded us that inventive military minds are constantly on the lookout for novel ways to kill people. Hundreds of thousands of scientists and engineers who work in defence labs or under defence contracts keep the military plugged in to ideas at the cutting edge of science, including the hottest area of biology — molecular genetics.

The Holy Grail of molecular genetics has become the elucidation of the entire sequence of letters in the DNA of a human cell early in the next millennium. Once determined, this base sequence will be like an immense standard reference book against which comparisons of any human group can be made relatively easily. Thus, it will be of interest to scan DNA sequences for similarities among patients with cancer, heart disease or mental defects in a search for clues that might lead to a treatment or cure.

But this kind of scientific knowledge also has the potential to be detrimental to human beings, especially with the keen attention of military minds. Consider this. The war in Vietnam and skirmishes in the Middle East, South America and Africa have pitted different ethnic groups against each other. Rushton's evolutionary notions about race and intelligence have revived the long-discarded idea of the biological significance of race in human affairs. Geneticists have known for decades that there are ethnic diseases, hereditary conditions that are more common among certain ethnic groups. For example, Tay-Sachs disease is found predominantly among Ashkenazi Jews, sickle cell anemia occurs at a very high frequency among black people, thalassemia is most common in southern Mediterranean people and cystic fibrosis is highest in the North American white population.

In 1970, a Swedish geneticist, Karl Larsen, published an article in the December issue of *Military Review* entitled "Ethnic Weapons." (It may just be a coincidence that the U.S. was then involved in a war with Vietnam and had fought in Korea just over a decade before.) In the article, Larsen pointed out that in addition to differences in diseases, ethnic groups may vary in their resistance or sensitivity to a number of chemicals. For example, Asians, Inuit and Amerindians who are closely related evolutionarily carry genes for digestion of certain components of milk. Those genes are turned off in adults. Thus, consumption of large amounts of milk by adults of mongolian background is liable to induce diarrhea and cramps.

If enough of such differences between ethnic groups could be catalogued, Larsen suggested, it should be possible to construct a cocktail of compounds that would be rigidly selective in its ethnic target. Larsen wrote his article long before the remarkable developments in biotechnology had become a reality and couldn't have anticipated what is now possible.

Once the complete sequence of human DNA is known, it will be very easy to search for base sequences unique to each ethnic group. Those characteristic ethnic sequences could become the target of ethnic weapons. A powerful technical tool now used by molecular biologists is a molecular probe, which can search out and attach to very specific base sequences within the DNA blueprint. Already such probes can be encapsulated within cell or viral membranes that are like magic bullets and can deliver their contents to specific targets. Ethnic probes could be chemically attached to compounds that induce severe sickness or death. Molecular biology provides all the makings for such cheap and highly effective weapons.

In order to avoid such grotesque abuse of their work, scientists must acknowledge the darker potential of their ideas, openly discuss the possible hazards, take strong public stands against proposals like ethnic weapons and press for an international ban on them. Public knowledge and revulsion are the best insurance against the development of such weapons.

Is It Dead?

*T*he nuclear arsenal remains the greatest single threat to survival of all life on Earth. While the Gorbachev initiatives have electrified the world and infused all peace groups with hope, there remain powerful vested interests from the "military industrial complex" to maintain the status quo. For the sake of our children, we have to keep demonstrating our opposition to nuclear weapons and push for a complete ban on all such weapons around the world.

It's remarkable how quickly children become aware of the fact of death. When my daughter Sarika was two years old, she knew the word *dead* and was constantly pointing to motionless insects, worms and fish and asking, "Is it dead?" And I would catch her squashing bugs and waiting to see whether they moved.

Seven years ago when my daughter Severn was three, her grandparents' beloved dog, Pasha, died. Then her great-aunt Myrtle passed away. She understood that death meant gone forever, but in one of those poignant moments that every parent would like to shield a child from, she suddenly turned to my wife and asked, "Am I going to die too?" It still hurts to recall that moment of insight, for when Tara said, "Yes," Severn wept inconsolably for hours that day — she had confronted that tragic reality of our species, the awareness of our own mortality.

One of the prices of our self-consciousness and intellectual capacity has been the recognition that death is the one inescapable consequence of living. How ironic that even the wealthiest and most powerful are all still ruled by this same biological event.

Of course there have been many elaborate ways that we have tried to create an illusion of escaping death. Many religions have provided a comforting alternative to the idea of eternal nothingness at the end of life. We also find reassurance in the idea that we "live on" biologically through our children or in the form of something we have done or created.

Modern science and medicine have provided a very powerful illusion that death may be conquered. With drugs, surgery, mechanical replacements, doctors bring people back from the brink of death, and often the achievements are stunning. But to my knowledge, no doctor has ever beaten death, only postponed it. Nevertheless, there is an underlying notion that aging and death are somehow unnatural and susceptible to understanding and eventually control by medical science. This medical battle with death is not much different from the attempts by ancient Egyptians to reach the other world; we don't want to die and will use whatever means are at our disposal to stave it off.

The irony of the twentieth century is that in a time when there have been astonishing advances made in prolonging life and increasing its quality, the same creative forces of science have generated the means of putting an end to it all. As Yale psychiatrist Robert Jay Lifton has movingly discussed, we have always found some kind of solace in the idea of living on in some way after our physical deaths. But the possibility of nuclear war puts an end to all that and even more. We have the idea of "nature" as eternal and endlessly self-renewing. If nuclear war can threaten all life on the planet, then not only do we face the extinction of our species but the death of nature itself.

Lifton recalls that shortly after the bomb was dropped over Hiroshima, a terrifying rumour spread through the country that trees and grass would never be able to grow there again. This symbolized the destruction of nature and was even more horrifying than the number of human lives lost.

Some Western religions speak of the final day of reckoning, Armageddon, which could be a metaphor for destruction by nuclear war. But in the Biblical version, it is God, not human beings, who sets those forces loose. It is horrifying to think that *we* now threaten not only our own survival and all notions of immortality but even nature itself.

Of course, nuclear annihilation is not the only way we have put the planet at risk, just the most obvious and immediate. We are using technology in the most crude but powerful ways to tear at the web of life. Nuclear weapons are simply the most powerful symbol of our destructive capacity. We think we are the only species for whom symbols matter. Can we respond to this one in time?

Severn has already been on many peace marches, and I hope that by seeing her parents trying to do something for peace she will grow up feeling that there is something that she as an individual can do. But I dread the day when she asks, "Daddy, why are we so afraid of nuclear bombs?"

WARRING SIBLINGS: ECONOMICS AND ECOLOGY

Introduction

*F*or most of human existence, our species lived off the abundance of the biosphere. When we were hunter-gatherers living in small family groups with simple technology, our impact on the surroundings was slight, and nature appeared to be vast and replenishing. People evolved elaborate means of trading and bartering for goods of use or value.

Human beings have a remarkable ability to place value on objects for esthetic or utilitarian reasons. Our ability to use abstract ideas or concepts, to cling to symbols as representatives of other objects led to the development of *money* as an abstraction of goods and services. As societies discovered agriculture, permanent settlements led to the concept of ownership of property. As populations increased in size and complexity, there was a differentiation of social roles and functions and money became an effective medium of exchange.

Today, global economics has become the preoccupation of every national government. In the industrialized countries, economic growth has become the yardstick to measure government performance. Indeed, economic growth and progress have become interchangeable terms. But faith in the necessity for such growth has blinded us to the hazards that result from it.

The New Math

*T*he greatest shortcoming of the human race is our inability to understand the exponential function." That is the melodramatic opener in a lecture by University of Colorado physicist Albert Bartlett.

You may think he's a typical dreamy academic who thinks that his tiny area of expertise is the most critical in the world, but having heard his entire speech, I'm inclined to agree with him.

You may well ask, what is an exponential function? It is simply a mathematical description of anything that grows steadily over a given period of time. There is probably nothing more sacred to economists than the notion that for a country to be economically healthy, there must be a steady, annual increase in GNP — that's exponential growth. Inflation, our expanding use of air, land, water, and energy this century are also examples of expotential growth. The world's population and our production of pollution are expanding exponentially.

The problem, according to Bartlett, is that we wouldn't bat an eye at a newspaper report stating that inflation (or whatever) has averaged 6.3 per cent. If, on the other hand, it was pointed out that at that rate the cost of everything would double in eleven years, we would pay attention.

Anything growing exponentially will have a doubling time during which it will become twice as large. That can be easily calculated by simply dividing seventy by the per cent growth per unit time. So with four-per-cent annual growth, doubling time is $70 \div 4 = 17.5$ years; for five per cent, it's $70 \div 5 = 14$, for six per cent, it's 11.5 and so on.

For the consumer, the consequences of exponential growth are most striking in the effect on prices. If you take seven-per-cent inflation over a single lifetime of sixty years, the results are devastating: a 55 cent litre of gas becomes $35.20, a $4 movie goes to $256, $30 worth of groceries reaches $1,920, a $200 suit, $12,800, an $8,000 car, $500,000 and a $90,000 home, $5,760,000! When old-timers think back on what once could be bought for a penny or a nickel, they're reflecting on the changes brought about by exponential growth.

Remember the fable of the man who did a favour for the king and for his reward asked only the number of grains of wheat that would be obtained by starting with one grain on square one of a chessboard and doubling it at each additional square? That is exponential growth. On the sixty-fourth square, the number of grains is equal to 2 x 2, 63 times and the total number of grains on the chessboard equals 500 times the 1976 worldwide harvest of wheat!

Even more stunning is the fact that the number of seeds on any given square is greater than the sum of the seeds on all previous squares! That means this century, as our use of oil, for example, has steadily increased, each time total use has doubled, we have required more oil than had been used throughout all history up to that point.

We would do well to remember that everything in the universe including itself is finite. Nothing within it continues to undergo exponential growth indefinitely; as Bartlett says, "In all systems, growth is a short-term, transient phenomenon."

If you look at the history of humankind on this planet, it is only in this century that growth has become such an obvious part of life. For virtually all our species' history, change was imperceptible, our impact on the planet was slight. On a graph of our numbers, use of food, air, water, soil, the speed and distance of travel and communication, the curves are virtually flat for ninety-nine per cent of our history. They only begin to turn up in the past century, and then in our lifetime, through exponential growth, they leap off the page.

It took all our history to reach a population of a billion people on the planet. But then in a century and a half, we doubled twice to four billion. Population increases in the industrialized countries have dropped dramatically, but there are parts of the world with doubling times of twenty to thirty years. Already impoverished, they will need twice as much food even though there will be no new land for agriculture and ocean yields have already peaked!

The curves for our use of oil and increased use of coal are every bit as sobering. We should be grateful to OPEC for confronting us with our profligate use of energy (and we should regret its current collapse). How can we possibly speak of a glut of something that is finite and nonrenewable? We are going to run out of oil during the lifetimes of our children. Right now we are riding on the resources of Third World countries, and if we had to live completely on our own, conventional oil would run out soon after the turn of this century.

Our dilemma is starkly illustrated by imagining a test tube with bacterial medium in it. At 11:00 we introduce one bacterial cell with a doubling time of one minute. So at 11:00 there is one cell, at 11:01 there are two, at 11:02 there are four, and so on until at 12:00, the tube is full. Then when is the tube half full? The answer, of course, is at 11:59. If you were a bacterium, when would you become aware that there was a space (or population) problem? At 11:58, the tube would be one-quarter full, at 11:47, one-eighth full and so on. If a bacterium were to say to its mates at 11:55, "I think we've got a space problem," he'd be laughed out of the tube — any sensible bacterium could see it was ninety-seven-per-cent empty and they had been around for fifty-five minutes! Yet they'd be only five minutes away from filling it up.

Suppose bacteria are like people and only at 11:59 do they wake up to the impending crisis. So they desperately throw money at their bacterial scientists. And lo and behold, those scientists deliver — in less than a minute, they invent three test tubes of food.

That is a phenomenal find — three times the existing supply! (Can you imagine how reassured we'd be if we made such an oil find?) How much time would that buy? At 12:00, the first tube would be full, at 12:01, the second would be filled and at 12:02, all four would be packed! Quadrupling the amount of food and space would only buy two more minutes of time if growth continues at the same rate. No amount of scientific effort will be able to add more than a few percentage points of increase in food production, land and resources, and many scientists believe we are well past fifty-five minutes.

Within the lifetimes of our children, the entire future of our species will have been decided. The curves now going straight up will have to level out and turn downward. If we don't do it deliberately, Nature will impose the final control.

It's strange how we think that we have to have greater consumption and growth to maintain our "quality of life." So today, to speak of aiming for zero growth or even negative growth is seen as antediluvian, a wish to return to outhouses and dirt floors. Yet in my childhood in the 1940s we lived with far less of everything (our family of six lived on less than $1,500 a year) and our lives were rich and full. We had central heating, running water and electricity.

It is nonsense to think that our current quality of life cannot be cut back; it must. Nor will it mean a reversion to the Dark Ages — we already consume at levels far beyond what is needed. The great survival attribute of our species was the ability to project the consequences of today's actions into the future. We took steps to avoid danger and that's how we got here. Yet now we seem unable to do that in the face of some very simple mathematics, the consequences of our worship of exponential growth.

The Ecosystem As Capital

Our failure to recognize the impossibility of maintaining exponential growth in consumption and the economy is creating a global crisis of catastrophic proportions.

In the summer of 1988, Toronto was host to both the Economic Summit and a conference on the Changing Atmosphere. Prime Minister Mulroney attended and made pronouncements at both. At the Economic Summit, he reiterated the importance of maintaining economic growth and at the atmosphere meeting, he urged action to avoid further destruction of the atmosphere. And he acted as if there was no connection between them! We continue to allow the same old concerns for growth, jobs and profit to determine the political and social priorities of government. Leaders of all major parties are hemmed in by pressure groups, ignorance and their personal value systems and seem to have neither the time nor the inclination to rise above the immediate exigencies of political survival and long-held political truths. Political "vision" seldom extends beyond the interval between elections, yet the environmental crisis must be seen on a longer scale.

It is *economics* that now preoccupies the media and politics. Maintaining growth in the economy by carving out a presence in a global economic community has become the raison d'être of almost every government in the world. Implicit in any economic system are arbitrary and irrational human values about work, profit and goals. They alone render economic systems opaque to prediction and hence beyond "management." But today the magnitude of the global trade and monetary system has taken us into new dimensions of complexity for which there is no historical precedent — we simply have no idea where we are heading.

Global economics must be exposed for what it is — a complete *perversion*. To begin with, economics is a chauvinistic invention, a human creation based on a definition of value solely by the criterion of utility to our species. As long as we can see a use for something and hence can realize a profit from it, it has economic worth. Yet it is the *ecosystem* that is the fundamental "capital" on which all life depends. Financial leaders manipulate the monetary system for immediate profit with little regard for environmental or human consequences. The current climate of laissez-faire economics in which the marketplace and private sector are being released from government constraints only ensures greater environmental depredation. We are only one species out of perhaps *thirty million*, and however much we may think we are outside nature and control it, as biological beings we remain dependent on clean air, water and soil like all other organisms. Economics has no ecological foundation because it dismisses air, water and soil bio-diversity as limitless "externalities" shared globally.

Our preoccupation with profit deflects us from taking effective action on such issues as the greenhouse effect, acid rain, salmon depletion, forest destruction or contamination of fresh water in an ecologically sensible way. Countries like Canada and Australia whose natural re-sources are the envy of most people on the globe squander their natural treasures in a rush to maximize profit. Even though we have barely begun to understand the scope of diversity of life on Earth and the complexity of atmosphere, oceans and soil, the drive for profit subsumes any concern for the long-term implications of our ignorance. These days when entire ecosystems are destroyed by clearcut logging, hydroelectric dams, farming or urban sprawl, we offer *money* to the dispossessed as if cash can compensate for things that are unique and irreplaceable.

Modern economics is perverted by an addiction to military spend-ing. Each year, over a *trillion* (U.S.) dollars are spent worldwide for defence, including the manufacture, sales and use of machines of death. (That's about $20 million every *second!*) The weapons trade consumes scarce resources but generates enormous profits for major industries and nations. However, it is the diversion of scientific creativity by the military away from socially and environmentally useful areas that is the greatest perversion.

Military research and development now consumes more than seventy-one per cent of all U.S. R & D while only nine per cent is spent on health research. "The U.S. government devotes well over twice as much research money to weapons as to all other research needs combined — including energy, health, education and food — and the Soviet pattern is believed to be similar." (R. T. Sivard in *World Military and Social Expenditures, 1987-88*).

Global economics is perverted because it impoverishes much of the Third World by seducing their people with the blandishments of technological "progress." High-tech weaponry, disposable goods, highly mechanized agriculture and substitutes for mother's milk have had devastating social and economic effects on less developed countries. To pay, Third World countries mortgage their future by selling off irreplaceable capital — their natural resources. Brazil, for example, has teetered on the brink of economic collapse for years. In order to keep up with a massive international debt, the country is destroying a unique world treasure, its rain forest, the richest ecosystem on Earth that is home to much of the biological diversity of the planet and affects the atmosphere, soil, water cycles and climate in ways we don't understand. It is criminal to destroy those forests merely to *service the interest* on international debt, for when they are gone, Brazil will still be mired in debt.

If our candidates for office make claims to genuine vision and concern for the future, they must be made to answer *profound* questions about the interrelationship between the environment and economics. For the sake of our children, we cannot afford to go on with business as usual.

On Our Own

*T*he February 22, 1988, cover story of *Newsweek* entitled "The Pacific Century: Is America in Decline?" raised the provocative suggestion that just as the United States has dominated the global marketplace in the twentieth century, so Asian countries may represent the economic force of the next century. This is highly relevant for Canada, as academics, politicians and journalists ponder the disturbing possibility that the Free Trade pact with the U.S. has tied us to a fading economic power.

The performance record and economic statistics of Japan and other Asian newcomers — South Korea, Taiwan, Singapore, Hong Kong and even China — have impressed the world and seem to portend the future direction for aspiring countries like Canada. But is there much difference in the economic arguments of those who favour closer ties with these Asian countries and those who want Free Trade with the U.S.? They both assume the absolute sanctity of continued *growth* and argue that without steady economic development, Canada will become an impoverished nation.

Let's indulge in the following exercise that the eminent American economist Kenneth Boulding once suggested to me. Try to imagine the consequences for this country if we woke up tomorrow to find that the rest of the world had disappeared, leaving only Canada surrounded by two hundred kilometres of sea. What would happen to us? Of course, there would be considerable upheavals brought about by this new state, but we wouldn't starve — we are one of the few breadbaskets in the world and our marine resources would still be vast. Even though we are a northern nation, we wouldn't freeze — we have a couple of decades of readily accessible oil, much more gas and a huge quantity of Alberta Tar Sands. The fact is, we have a disproportionate share of the planet's fresh water, clean air, minerals, trees and much more — we are gifted with bountiful resources. And we have a highly educated population that can produce everything our society ultimately needs. If *we* are regarded as economically weak and poor, there must be something drastically wrong with the way we are being evaluated.

Now let us extend this little thought exercise and imagine the consequences if the *Japanese* woke up and found that all that remained was their country and the surrounding two hundred kilometres of sea. Unlike us, they would immediately find themselves thrown into a desperate struggle for survival. Deprived of all but a minuscule amount of energy, Japan would find its highly efficient use of energy would be of little consequence because the nation is completely dependent on foreign countries for oil, coal and gas. Japan could never again achieve the productivity it now has.

And what about food? The Japanese sit at the very top of the food chain — they exchange their high technology for the rest of the planet's raw resources or use it to plunder the world's oceans. The water and air around Japan are too polluted and overused to yield enough food to satisfy the people's demands. Japan is a major predator of this planet's resources and manages to pull it off because of a global economy that places immense value on high-quality manufactured products that *are not essential for survival*. The Japanese have taken advantage of a global market and created a demand for their products. But compared to Canadians, survival for the Japanese is a far more precarious proposition.

Last year, I teased Kenneth Boulding by saying that economics is so limited that it has no equation or factor for the value of an entire species. So to an economist I argued, *extinction* is unimportant or valueless. He retorted, "I contribute $25 a year to the Save the Whales Fund. That's about what they're worth to me and if I can convince enough people to kick in the same amount, we'll be able to save them." It was a clever answer, but at the very moment when species are being exterminated daily by short-term imperatives of human greed, opportunism or individual survival, Boulding's reply was too glib.

We pride ourselves on being a species that cares for its own and other species, but we have invented a system that works against those very values. Debating whether to copy Japan or the U.S. merely ignores the underlying disease that afflicts this planet — the pernicious assumptions of the global marketplace and the equation of growth, profit and consumption with progress. Until economists factor a sense of ecological place and the value of living in balance into their equations, it won't matter which economic star we choose to ally ourselves with.

Ecologists and Economists Unite!

*T*he words *ecology* and *economics* derive from the same Greek word, *oikos*, meaning household or home. So ecology (*logos* meaning *study*) is the study of home, and economics (*nomics* meaning *management*) is home management. These two fields should be companion disciplines, and yet with few exceptions there is little communication between them.

Even though the fundamental capital that all countries depend on is the natural world, modern economics makes no ecological sense. When a businessperson evaluates a forest, for example, that ecosystem is transformed into "board feet" or "cubic metres" that can then be plugged into the proper equations. Other factors — costs of surveying and road and bridge building, labour, reforestation, market demand and profit — are weighed before deciding whether a forest is worth logging. But considerations of the worth of *not* touching the forest are dismissed as "externalities" to economic calculations.

For most of human existence, we could get away with thoughtless extraction of "resources" from the environment because of the abundance and diversity within the natural world. Our numbers were small and technology simple and powered by human and animal muscle power. (That was still enough to build the pyramids and the Great Wall of China and to transform a number of forests into deserts.)

The invention of machines and exploitation of cheap, plentiful fossil fuels created a sudden and massive increase in technological muscle power that has had enormous ecological repercussions. Today our species alone has the power to affect the other thirty million on the planet. Almost overnight, we can destroy entire ecosystems. But conditioned by the long-standing resilience of nature, we've continued to act as if it is virtually limitless, and this behaviour is reflected in economic systems.

The planet is being ravaged for economic returns. But any farsighted economist must recognize that there are "services" performed by nature itself that have to be factored into the economic equations. So let's start by remembering that we are animals. As biological beings, we must have clean air, water and food for our sustenance and health. The biological world around us has assured us of that. In the past, pollution by our fires, leftovers and body wastes were recycled by other organisms. Today, the sheer magnitude, variety and novelty of our technological excreta preclude that.

The great forests of the world have served to modulate the water cycles of the planet, absorbing rain and transpiring it into the air or releasing it into the ground. Thus, ground water, erosion, flooding, landslides and weather are affected directly by forests. Forests also absorb carbon dioxide while releasing oxygen, thereby conditioning the air we breathe and the upper atmosphere that affects climate. Old-growth forests maintain a high degree of biodiversity on which long-term ecological stability depends. All these "services" continue to be performed as long as the trees are left standing, yet none of them are cost accounted before a forest is cut down.

There are other benefits of natural systems for humans that are seldom assessed economically. The most obvious is the enormous human capacity to discover and then exploit other species. Many of our most powerful medicines still are biologically based. The vast pharmacopeia of traditional medicines and yet-to-be discovered plants in tropical rain forests promise far greater returns than the much ballyhooed biotechnology.

Throughout history, people have used perhaps 7,000 kinds of plants for food, yet there are at least 75,000 edible plants, many superior to ones we currently use. Only about 150 have been grown commercially, while human nutrition today is based on only twenty or so major crops! (Of these, three grass species, rice, corn and wheat, are the most important.) There are also very real esthetic, spiritual and philosophical values for nature that are never reckoned in any economic model. If the forestry industry can be compensated when forests are preserved for parks, why shouldn't society be compensated for the potential lost when trees are logged?

Harvard University economist Julian Simon complains that ecological critiques perpetuate a myth of scarcity and dwindling resources. Simon clearly states the absurd faith held by most economists: "There is no reason why human resourcefulness and enterprise cannot forever continue to respond to impending shortages and existing problems with new expedients that, after an adjustment period, leave us better off than before the problem arose."

Lester Brown, president of the Worldwatch Institute, counters: "The lack of ecological awareness has contributed to some of the shortcomings in economic analysis and policy formulations." Brown points to fisheries, forests, grasslands and croplands as key areas for the global economy:

The condition of the economy and these biological systems cannot be separated. As the global economy expands, pressures on the earth's biological systems are mounting. In large areas of the world, human claims on these systems are reaching an unsustainable level, a point where their productivity is being impaired. When that happens, fisheries collapse, forests disappear, grasslands are converted into barren wastelands and croplands deteriorate along with quality of air, water and other life-support resources.

Economists cannot afford any longer to ignore their companion discipline of ecology.

Industrial Doublespeak

Why can't we see the flaws in economic thinking? One reason is that our values and beliefs are shaped by advertising. Slick magazine ads and dazzling television commercials blast a steady stream of facts and come-ons at the consumer. People are justifiably suspicious of these messages, but there is no doubt that multimillion-dollar ad campaigns have an effect. Our consumption patterns bear that out. But there is a corrosive effect of the incessant barrage of self-serving material on the very meaning of *facts* and *truth*.

This brings to mind one of my favourite books, *Ishi*, by Theodora Kroeber, which documents the life of one of the last Stone Age people in the U.S. near the turn of the century. Relentlessly hunted down by California settlers and gold seekers, all Ishi's people and family have been murdered by the time he is captured. This "wild man" is rescued by anthropologists and brought to Berkeley where he learns English and holds a job until his death. San Franciscans ridicule Ishi because he believes anything they tell him. They call him a "dumb savage" because they can cheat him so easily. What those "civilized" city folk don't know is that in Ishi's culture, lying was unheard of — they didn't even have a word for it!

Ishi gives us a chance to see ourselves from a different perspective. Implicit in the way he's treated is the sense that it's fine to cheat and lie if the victim is gullible. I think of that as I watch the predictable sequence of response by various North American industries to attacks from the public. Initially, the industry denies the validity of any criticism while belittling the opponents as overly emotional, unrealistic romantics — people who lack credibility and are against progress. The industry usually quibbles over petty details and steadfastly refuses to acknowledge the legitimacy of their critics' major concerns. If the pressure builds, industries often portray themselves as beleaguered victims of a well-organized, heavily bankrolled outside group and warn that there could be a loss of jobs and serious damage to the economy if the opposition succeeds. Pushed to the wall, the industries launch massive, expensive public relations campaigns with ads and commercials deliberately designed to trivialize the important questions.

Perhaps the most blatant example of the pattern of response to public criticism has been the tobacco industry. The link between smoking, heart and lung disease and mortality is *proved* beyond doubt, yet the industry still refuses to acknowledge that a single death can be attributed to smoking. This denial is a slur on scientific and medical experts and on science itself. Currently, tobacco companies refuse to confront the hard facts about the undeniable health hazards to smokers and their fetuses, the highly addictive nature of tobacco and the enormous economic and social costs of smoking. Instead, they have begun a multimillion-dollar ad campaign to fight for their right to advertise on the basis of *free speech*! It's simply not credible from an industry that has such a poor record of open discussion of facts.

The nuclear industry provides another example of public concern being handled less than forthrightly. The industry has been subsidized by billions of dollars from the Canadian taxpayer. In the U.S., the cleanup bill for Three Mile Island has already reached $1 billion, and no end is in sight. In the event of a major accident, Ontario Hydro, Canada's main generator of nuclear power, is exempt from costs above $75 million, even as their experts proclaim the probability of a serious accident is vanishingly small. If it's so safe, why the upper limit on liability? Ontario Hydro has been excused from carrying out environmental-impact assessments on any of its nuclear generating stations or the tritium removal facility at Darlington and has never been required to spell out how highly toxic long-lived products can be safely disposed of.

Three Mile Island and Chernobyl have badly damaged the credibility of the nuclear industry, yet its supporters continue to avoid confronting the issues of public concern while nitpicking "facts," disparaging the expertise of its critics and proclaiming the credibility of its experts. And now, the Canadian Nuclear Association, which is heavily supported by Atomic Energy of Canada Ltd., which in turn is subsidized by our tax dollars, has initiated a multimillion-dollar ad campaign to convince the Canadian public of the safety and necessity of nuclear power.

The Canadian forestry industry is an important contributor to the economy and has long portrayed forests as a "renewable" resource. Canadians value wilderness and a clean environment and have become aware of the horrifying speed with which it is disappearing. Natives, environmentalists and critics of the economics of the forestry business have pointed to the industry's failure to reforest, the destructive effects of clearcut logging, the enormous waste, the short-sighted drive to maximize immediate profit and the failure to address spiritual and esthetic values of old-growth forests.

So now forestry giants have started a slick and expensive ad campaign to portray themselves as a caring industry that is preserving and restoring wilderness in a responsible way. It is a blatant example of misinformation through advertising, which ecologist/journalist Bristol Foster has termed "ecoporn."

If critics had even a fraction of the money being spent on PR by these industries, society and truth would be far better served.

Infoglut and Its Consequences

Access to more information is supposed to help us perceive reality and deal with the big issues that confront us. And by any criterion we want to apply, more information is available to us than at any other time in human history. Information overload, however, may create problems as severe as a lack of information.

Books, articles and television programs herald the arrival of the

Information Age. Telecommunications and computers form a global network that has become a key to economies of the world. We are the global village foreseen by the late communications wizard Marshall McLuhan and trumpeted by such futurist authors as Alvin Toffler.

When I began my career in television in 1962 it was my perception that science, when applied by industry, medicine and the military, was by far the most powerful factor shaping our lives. Consequently, I felt the public needed more information about science to make better decisions about important issues in their lives. I was completely wrong.

By every standard we wish to apply, Canadians already have access to more information than ever before, but we still seem incapable of coping with the immense issues of nuclear war, pollution, species extinction and so on. The problem is that we are inundated with information. Most of it is worthless to the average person, but we have no idea how to wade through the morass and separate the meaningful from the trivial.

Scientists refer to a situation such as this as having a "low ratio of meaningful signal to background noise." We have to ask critical questions about the nature of this information: What is it? Who provides it? What does it mean?

Today, we have specialists who tell us what we think. They are called pollsters. They sample public opinion on all kinds of issues for provincial and even municipal offices. But what do they tell us? During the lead-up to the Quebec referendum on the province's future association with the rest of Canada, the Parti Québécois agonized for months over the wording of the referendum question because they surmised that phrasing would strongly influence the outcome. The questions posed by pollsters are not simply objective searches for information. The way they are asked influences the kind of responses that will be generated.

In any poll, the greater the number of response choices for any question, the more difficult the analysis. Hence, questions with yes-no answers are preferred. But the fewer the choices, the less meaningful the answers. Surely when we are asked whether we favour Free Trade with the United States, membership in the North Atlantic Treaty Organization (NATO), participation in the Strategic Defense Initiative or strong environmental legislation, simple yes-no answers reveal almost nothing of the complexity of either the issues or the thought behind the responses.

Even increasing the number of categories to "definitely yes," "yes," "perhaps yes," "perhaps no," doesn't add very much. Usually the answers are reduced to mere percentages or pooled categories (all the yesses — "definitely" and "possibly" are simply lumped as "favouring"). We vest a great deal of value in numbers that cannot inform us about the variety of responses, the concerns and qualifications that shape our answers.

Nothing better illustrates the relative meaninglessness of information when it is simply packaged as a number than the distribution of scores on IQ tests. Forget the question of what IQ tests actually tell us about intelligence, because that is a matter of great debate. It is a fact that when a number of people take an IQ test, their scores will fall in a distribution that approximates a bell shape. People fall on either side of 100, and the farther away from that mean score, the fewer there are. Those who score above 120 represent less than ten per cent of the population.

In a city the size of Montreal, there may be 100,000 or more people who would score higher than 120. But so what? What does an IQ score tell us about the enormous human variability that resides in all the people who share that number? They will have just as many stupid, mean, avaricious, generous, caring individuals as any other group along that curve.

By reducing people to mere ciphers, we give the appearance of scientific objectivity, but there is little real informational content. Most of these numbers tell us nothing abut the complexities that make issues important; they hide human values, fears or hopes, social injustices or impediments to progress. Today, much of what the information purveyors do is collect massive amounts of data, condense it and subjectively decide what is important.

I once gave a talk to executives in the electrical industry. One complained of the proliferation of books, magazines and articles in his area. When I asked him how he dealt with them, he simply snorted and replied, "Hell, I don't have time to wade through the stuff. I tell my people to give everything to me on two pages or less. I prefer graphs with colors." It didn't surprise me, but it gets rather scary to think that decisions involving thousands of employees or millions of dollars may be made on the basis of a two-page summary of a one-hundred-page report.

Consider cabinet members in federal and provincial governments. They work under enormous pressure and have no time to reflect on long-term implications or to read books and scholarly articles. They depend on their aides to summarize material, but what is lost in the reduction?

Daily we are assaulted with commercials, articles and reports proclaiming the wonders of the Information Age. But we don't spend much time asking what it's all worth. We get our information today principally from television. We watch TV in large blocks of time — it is the main way we learn about ourselves and the world. And it's easy to see why: television is an almost effortless way of being informed, it's pleasant and pictures make a powerful impression.

But if we are to be well informed, we must understand the medium, its limitations and modus operandi. We must realize that television does not reflect reality. Television reports are artificial, just as movies are. We don't have zoom lenses in our eyeballs; we don't live in a world where we can "cut" from one place to another or from one time to another. Yet that is what television does, and our brains fill in all the spaces. Try watching a report of a single event, such as a pole vaulter clearing the bar or a theatrical performance. It can be an illusion created by the juxtaposition of several cameras or repetitions edited together into an apparently single continuous piece.

News reports themselves are fabrications. From the decision as to what is newsworthy to the availability of a crew, the cost of doing the report and deadlines for editing and broadcast, stories sometimes are made or broken for reasons that have nothing to do with the significance of the events themselves. Once something is judged worthwhile (and that is often arbitrary and subjective), the item's length may be determined as much by whether an interview subject is articulate, or by the kinds of supporting visual material available, as by the inherent merits of the story.

And most important, if the researchers, writers and reporters are upper-middle-class white males, chances are their unconscious social and cultural biases and values will colour the way the report is finally presented. I say all this not to denigrate the medium, but to point out that there is no such thing as objective reporting on television and people should understand that as they watch.

When I began my career in television in 1962, it was my conceit to think that through quality programming I could educate the public and raise its general awareness of the kinds of issues I was interested in. A quarter of a century later, *I* have been educated about the severe constraints in this endeavour. The problem is one of sheer *volume* of information: there's so much to absorb and we no longer have a single coherent framework of values and perspectives to use to assess the information.

Change is what characterizes our lives — all about us, change is the one dependable feature of our social landscape. And because of rapid change and turnover, we acquire bits of information and quickly sift through them, retaining only those that happen to strike a chord. So often we repeat a "fact" that we justify with "I read somewhere that . . . " or "I saw on TV that . . . " without regard to the quality of the source. If the *National Enquirer* is viewed as being equally as credible as *The New York Times,* then information becomes totally devalued.

I am constantly surprised when in letters, phone calls or personal encounters I am credited with all kinds of reports from programs I had nothing to do with. Often I am told, "I saw on your show . . . " but it is a story we have never covered. People do not watch TV with concentration; there are interruptions — phone calls, getting a snack, putting the children to bed. We tune in and out, watching when something interesting comes up and drifting off at other times. By the time we go to bed, the contents of a four-hour viewing block may be completely mixed up, and it is easy to assume that a snippet remembered from one show was actually seen on another.

We do it with other media as well, retaining little nuggets of information. This came home to me once when my wife, who teaches expository writing to science students, brought home a copy of the *National Enquirer* that she had used in a class exercise. I perused it with interest and found it was full of fantastic stories on medicine (woman inseminated by robot) and science (Soviets capture UFO crew). Three days later at work, I found myself mentioning that I'd read that there are toxic chemicals in the nipples of baby bottles. Midway through my recitation, I realized with a shock that I was repeating a story from the tabloid!

So while we have access to more information than ever in human history, "infoglut," rather than educating and informing us, can make life much more complex and difficult. If we are going to take advantage of information, we have to understand the media and their limitations.

We need to assess the source of information. We have to believe in our own ability to judge the pros and cons of controversial issues and to demand access to primary information. That is precisely what scientists do in their profession. They are skeptical of any new claims, demand to see the evidence for themselves and are confident enough to trust their own judgement about the worth of new claims. It's an attitude that the general public should find of great value in a time of informational overload.

THE ENVIRONMENT: THE SCOPE OF THE PROBLEM

Introduction

*I*n the mid 1950s, while I was returning from college in the U.S. to my hometown of London, Ontario, for the summer, I glanced down at the Niagara River as my train passed over it. Far below, I could see people on the banks flailing away at the water and yanking silver objects from the river as fast as they could cast. The fishermen were intercepting silver bass on a massive annual spawning run that had probably been taking place each year for millennia. Yet within a decade, those spring runs had disappeared.

Old fishermen on the east and west coasts of Canada reminisce about their early days when they caught huge quantities of large salmon, cod and lobsters such as younger fishermen have never seen. It wasn't long ago that we drank from the Great Lakes with confidence in the water's purity and relished fresh fruits and vegetables without concern about chemical contamination. Only a few decades ago, the quality of our water, soil and air were radically different, and there was an abundance and variety of wildlife that is now found only in the most remote parts of the country.

Within the lifetimes of our elder citizens, the planet has changed almost beyond recognition. Their childhood recollections are not simply the romantic musings of old folks for the good old days, but they constitute a living record of the cataclysmic degradation that has taken place around us in the span of a single human life.

My childhood dreams of experiencing the endless expanse of Amazon "jungle" were shattered by a visit this fall. The Brazilian rain forest is being burned, flooded and leveled with horrifying speed. If biologists are correct, virtually all wilderness on the planet will disappear within the next thirty years. This means that we are making decisions now that our children will live with and all future generations will inherit. Yet society seems locked into a single-minded pursuit of profit and growth.

What Is the Problem?

During the seventies and eighties, environmentalists were popularly thought of as the "crunchy granola" set, back-to-the-land hippies or Greenpeace "crazies." But suddenly, there has been a remarkable shift, and 1988 may have been the critical year.

Any report of the significant events of the year expresses the values and beliefs of its author, and this one is no exception. In 1988, preparation for a speech I would give at the thirtieth anniversary of my college graduation provided an opportunity to reflect on the vast changes that have taken place during my life. In the context of that half century, the truly significant events of the past year were the unmistakable signs of catastrophic change in the biosphere. All we have to do is look.

In 1988, more groups of indigenous people around the world solicited international support to stop destruction of their lands. In Sarawak, the Penan people, inspired by a remarkable Swiss man, are beginning to fight against the logging that is destroying their home to provide timber to Japan. In October, B.C. Chiefs Ruby Dunstan and Leonard Andrew flew to Wellington, New Zealand, to attend the annual meeting of shareholders of Fletcher-Challenge, a multinational corporation. There they stated their opposition to the company's plans to log the Stein Valley, a sacred area for the Lytton and Mount Currie Indians.

Indians in the Amazon rain forest are rapidly losing their way of life through the destruction of their land by projects such as massive hydroelectric dams and multinational iron smelters. In November, Chief Paiakon of the Kaiapo Indians travelled through Europe, Canada and the U.S. to raise money and generate opposition to World Bank and commercial bank loans to Brazil. Over 3,000 people in Toronto and 800 in Ottawa jammed halls to hear him speak and contributed more than $80,000 to his cause.

Across Canada, the Gitk'san-Wutsewatin people in B.C., the Lubicon in Alberta, the Teme-augama Anishnabi in Ontario and the Innu in Labrador, among others, are fighting to preserve their wilderness homes intact. We discover in such struggles that these people have a *spiritual* relationship with their land that is both real and beyond price.

Once again in 1988, there were intimations that something has gone wrong on the planet. An exceptional heat wave and dry spell through the centre of North America served as a preview of the consequences of global heating by the greenhouse effect. A devastating flood in Bangladesh had its origins in Nepal where deforested mountains no longer absorb the rains. Seals dying unaccountably in the North Sea became yet another symbol of the deteriorating biosphere. We were thus reminded that air, water and soil are not matters confined within state or national borders.

In 1988, for the first time, candidates for political office could not ignore the environment as an issue of public concern. In the federal election, the major political parties assumed postures on the environment that acknowledged its existence but did not propose profound solutions. Henceforth, the environment will be a central issue in every election.

In future, the conflict between economic priorities and environmental protection will have to be resolved. Thus, for example, serious solutions to the atmospheric degradation described in Toronto at the Atmospheric Conference can only be proposed *within* the context of the objectives of the Economic Summit. Increasingly, popular media like *The Globe and Mail* will have to reconcile the inconsistency of taking a progressive environmental stance while simultaneously supporting aggressive policies of growth and development in its business sections.

We can end our reflections on 1988 with the poignant images of three whales trapped by ice off Alaska. Around the world, people responded emotionally to the plight of the great leviathons and the heroic efforts of their international rescuers. But what was the basic message of this media event? It depicted a battle of human beings against the forces of nature and with human pluck, ingenuity and technological power, we seemed to triumph over nature. But we must not allow that exercise in compassion to blind us to the fact that it is the hubris implicit in our symbolic conquest of nature that remains the greatest threat to all other life forms today.

It Takes a Disaster

*I*f 1988 marked a major shift in attitude to the environment, why did it take us so long to recognize the signs? A first sight of
Los Angeles from the air imprints an indelible picture. Peering from
the plane window, we are greeted by an immense brown dome. As the
plane sinks through the thick air, one feels like a swimmer being sucked
helplessly into an alien environment. Remarkably, what looks like a
thick goo from air is not nearly as daunting at ground level. On my first
visit to L.A., a taxi driver announced that it was a "good day." I made
note of the fact that a good day meant you could see six blocks. Within
a day or two of arrival in Los Angeles, one forgets the initial sneezing
and watery eyes — adaptation is quite rapid. I have visited Cal Tech,
which is located in Pasadena, when the smog was so thick that visibility
was less than half a block! Yet people there merely shrugged and went
on with their daily business.

One of the striking results of wiping out all life through acidification
of lakes is the clarity of the water. A remarkable observation was made
in Scandinavia. Researchers found that a year or two after the initial
shock of realization that the lakes had been sterilized, cottagers along
the shores *preferred* it! Apparently people preferred swimming without
pesky weeds or slime while much of the noise from motorboats was
eliminated because there were no more fishermen to indulge in their
annoying habit of starting early.

People who have worked for years trying to arouse public support
against nuclear weapons often experience frustration with a common
reaction that members of the general public just "don't want to hear
any more about nuclear war." Indeed, the raw facts on the present level
of armament and the consequences of even a limited nuclear war are
so horrifying that thinking about it for any length of time threatens
one's sanity.

In all three cases — smog, acid rain and nuclear war — why is it
that people seem so tolerant of what is so obviously unacceptable? We
have a lot of names for it — apathy, burnout, denial, fatalism, indiffer-
ence, passive avoidance, learned neglect — but whatever the term, our

ability to blot out the reality of such problems prevents us from working on a strategy for change.

Perhaps there is a physiological explanation for our apparent psychological malleability. Neurobiologists can place an electrode in a nerve to measure the electrical response when that nerve is stimulated. But if the stimulus (for example, a flash of light on an organ of vision) is repeated in progressively shorter intervals, the nerve eventually stops responding. It is a phenomenon that neurophysiologists call *habituation*. We've all experienced this — when a loud bell goes off we are acutely sensitive to the noise but when it continues to ring, we begin to ignore it and as our attention moves to other things, we don't even notice it. That's habituation.

From an evolutionary standpoint, it's easy to see the reason for habituation. In the event of something catastrophic like a flood, a violent storm or a terrible accident, we have to register what's happening, but must not be overwhelmed by the immensity of it all. There has to be a mechanism to dampen the scale of our response so we can try to figure out ways to survive.

Evolution prepared us to respond to immediate crises but not for intangible possibilities that *may* happen far in the future. That's because we rapidly habituate so the intensity of the response to the threat can't be sustained. Habituation developed as a survival mechanism, but today it acts against us by desensitizing us to very real but *distant* perils.

Does this mean there's no hope? Once again neurophysiological findings may be instructive. There is another fascinating property of nerves called *sensitization*. An initial stimulus to a nerve elicits a standard response but a second one delivered in a different way can produce an even *greater* nerve reaction than the first. If you've ever walked down a path in the dark and been surprised when a bird suddenly flutters off, you know how heightened your sensitivity is in anticipation of another encounter with an animal. That's sensitization.

In the past years, the number of newspaper reports and magazine articles on environmental issues has steadily increased. The names immediately remind us of the stories — thalidomide, DDT, dioxin, PCB, Minamata, Bhopal, Three Mile Island, Challenger, Chernobyl, ozone, greenhouse effect. Each additional story is unexpected and induces a greater awareness of the inescapable global crisis. Remember the PCB spill on an Ontario highway near Kenora a few years ago? There was a brief flurry of stories and then it faded away. But then a PCB-laden cloud shrouded St-Basile-le-Grand and that time, it did not disappear from the public mind so readily. PCB has become a familiar term that now reminds us that we live in an ocean of man-made chemicals.

So while smog, acid rain and nuclear weapons may now be old hat, each addition to the growing list of technological and environmental disasters jars us out of the numbing effects of habituation into a higher level of sensitization that is the prerequisite to action. Today, there is a groundswell of public sentiment that something is wrong with the environment and no politician can ignore the need to take drastic and far-reaching action.

Going Down Under

*T*he current global crisis in the environment is the legacy of an attitude towards the great riches of the planet. The crisis is often hard to recognize in the society we've lived in all our lives. It may be easier to see in another place. For me, this has certainly been true; a case in point is an article I wrote in July 1988.

As I write this column in a biologist's paradise — Australia — a flock of brilliant red birds with pastel blue wings has just wheeled overhead and landed in a tree I can't identify. Nearly every insect, bird and flower I see is brand-new to me. A dive at the Great Barrier Reef was sublime.

In case you haven't noticed, this is the year of Australia's bicentenary, and the orgy of celebration signals an end to the "cultural cringe" from its convict past — today the cosmopolitan urbanite is as real as the rugged individualist in the outback.

Looking at the unique geological and evolutionary history that has created this continent also provides a very different perspective on Australia's celebration. The country began with the breakup of Gondwana, that fabled supercontinent, hundreds of million of years ago. After finally cutting loose from Antarctica about forty-five million years ago, Australia carried most of the animals and plants that would follow their own unique evolutionary path. The only mammals on the continent were marsupials and the evolutionarily older monotremes, the egg-laying platypus and echidna.

An incredible array of animals evolved — gigantic mammals, non-flying birds, lizards and snakes — that we know only through the fossil record. During the last great Ice Age, so much water was tied up in the Antarctic ice cap that the oceans dropped by some thirty-five metres and brought other lands within hailing distance.

At least 50,000 years ago, human beings first reached the continent and, just as happened when people advanced into North then South America, the large mammals and reptiles quickly disappeared. Human intelligence, aided by the use of fire and simple tools, was enough to outwit large animals and drive them to extinction. But very quickly, those ancestors of today's Aborigines achieved a balance with the incredible flora and fauna that were still there when Europeans arrived two centuries ago.

Australia's original inhabitants, like aboriginal people around the world, were deeply involved in the natural world physically and spiritually. The European arrivals brought something far more destructive to the natives than disease and alcohol — an obsession with ownership and money. Aborigines were simply added to the list of Australia's flora and fauna (and were only removed in 1967!) and exploited like the rest of the ecosystem.

Ecologically, the occupation of Australia by Europeans has been catastrophic. The statistics tell the story. Seventy-eight species of native plants are already extinct and two hundred more are endangered. Sixteen mammal species are gone (including wombats, wallabies and the Tasmanian tiger) and 106 of the remaining 204 are threatened (including the koala and the duck-billed platypus). Three bird species are extinct and eighty-six of 720 are threatened. Of 560 kinds of reptiles, 123 are threatened; of 160 amphibians, thirty-seven are in danger; of 200 freshwater fish, three are extinct and thirty-five threatened.

Equally as damaging as species loss has been the accidental or deliberate introduction of new species to the continent. The water hyacinth has been as lethally prolific here as in Canada and joins the prickly pear cactus, English sparrow, starling and European honeybee as disruptive newcomers. At the time of continental separation, only marsupial mammals (which rear their infants in pouches) were present in Australia, so they have flourished and diversified uniquely. Animals with placentas, those that give birth to more mature young, have often been deadly competitors who have wreaked havoc on their marsupial cousins. Of thirty-nine introduced species, twenty-seven are known to have established a foothold. Introduced species that manage to survive are free of the natural control mechanisms that exist where communities of organisms co-evolve over millennia. So in Australia, rabbits, pigs, rats, cats and dogs have been to the ecosystem what smallpox and measles were to aboriginal people.

The major human impact here has been the destruction of *habitat* for wildlife. Six million hectares of rain forest have already been cleared, leaving only about two million more. Sixty per cent of the coastal wetlands of New South Wales alone have already been cleared, drained and filled. Of Australia's 768 million hectares of land, only 2.2 per cent is forest! A whopping 65.2 per cent is already used for agriculture or grazing, and 25.8 per cent is either desert or belongs to natives.

But even the farmland is not being cared for properly. Per capita, Australia has three times as much degraded farmland as North America. That includes four million hectares of soil destroyed by salinity and seven million hectares by acidification. Fifty per cent of farmland is being degraded by soil erosion at an estimated cost of $3 billion annually. It may cost about $2 billion a year to halt the erosion, but the government is spending about $100 million.

Like Canadians, Australians look at marine organisms as limitless and free. So if foreign countries are willing to eat creatures Aussies don't find appetizing, they are happy to extract and sell them in vast quantities. (We do it in Canada with herring roe, geoducks, snow crabs and gooseneck barnacles, even though we know almost nothing about the basic biology of these species. Our "harvests" are little more than short-sighted mining operations.) In Australia, even though the fishing fleet has increased greatly over the past decade, annual catches of rock lobsters, scallops, prawns and tuna have been falling precipitously for five years. Any thoughtful person must know something is wrong, but the mindless reaping goes on.

Australian business and political leaders are determined to carve out a niche in the global high-tech economy, a path that will only increase the destructive demands on their environment. The Bicentenary should be an occasion for reflection on the implications of the past for the coming two hundred years. Modern Australia is an ecological disaster characterized by a squalid history of greed and ignorance. It will be a tragedy for the world if what makes this such a fabulous and unique continent goes the way of the passenger pigeon.

The Rape of the Amazon

*I*n my childhood, names like jaguar, anaconda and piranha summoned up thrilling images of fabulous creatures and natives hunting with blowguns in the Amazon. In October 1988, when I wrote this in Belem, where the Amazon River reaches the Pacific, I found a far different picture. There, as everywhere else on the planet, greed and short-sightedness fuel a rush for profit and security. The cost is the irreversible destruction of a unique treasure that is beyond price — the Amazonian forest and its inhabitants.

Tropical rain forests are being depleted faster than any other ecological area. While they only cover seven per cent of the world's land surface, they contain at least half of all species on the planet. And none is more rich and varied than the forests of the Amazon. Although few comprehensive studies have been made on its biodiversity, scientists know that the Amazon forests harbour at least 50,000 species of higher plants, an equal number of fungi, a fifth of all birds in the world, at least 3,000 species of fish (that's ten times the number in all of Europe) and untold millions of insects. A typical hectare of Amazon forest will contain 200 to 300 species of trees alone! The distribution of organisms within it is not uniform; they form localized clumps of plants and animals, many uniquely adapted to a restricted area, so large-scale destruction of the forest inevitably results in species extinction.

This great biomass is a major influence on the water cycle for the entire continent. The airstream originating above Africa sweeps across the Atlantic, picking up moisture that is dumped on Brazil's coastal forests. Twenty-five per cent of the rain never even hits the ground, but evaporates from the forest canopy. The rest is cycled through vegetation and animals below and eventually is transpired back into the air by plants. Thus, water cartwheels across the continent five to seven times until the winds hit the Andes and split north and south. Weather and climate are modulated through this complex cycle.

The scale of destruction of Brazil's forests is illustrated by the state of Rondonia at the upper reaches of the Amazon. Between 1975 and 1986, Rondonia's population exploded from 111,000 to over one million! (Try to imagine the disruptive consequences of a tenfold increase in people in a city or province in only ten years!) In 1975, almost 1,250 square kilometres of the state's forest were cleared. By 1982, this had grown to more than 10,000 and by late 1985 to about 17,000 square kilometres. That's over ten per cent of the state's forests, and the rate of deforestation continues to climb.

For months during the burn season, the air is filled with smoke, the weather is changed and planes can't land. Satellite photos reveal that Brazil is ablaze with as many as 7,000 fires visible at once. In 1975, only one per cent of all forest in Amazonia had been destroyed, but by 1980, four per cent was gone! Today, well over ten per cent is gone and the soaring rate of clearing puts the entire rain forest in imminent peril. And that's not all. Massive hydroelectric projects flood huge areas of uncut forest and drive out all inhabitants, human and nonhuman.

Brazilians believe Amazonia is "a land without people for people without land," so forest land is made available free — all you have to do is clear the trees. It's a cruel promise. In northern temperate forests, a thick layer of soil accumulates forest debris over thousands of years and contains over ninety-five per cent of the nutrients for forest productivity. In contrast, over ninety-five per cent of the nutrients of a tropical rain forest are in the vegetation. Forest litter is quickly recycled by microorganisms, animals and plants, so the nutrient layer of soil is extremely thin (less than five centimetres) and fragile. It's virtually useless for farming or raising cattle, yet the trees continue to be cleared and burned for that purpose.

The rain forest consumes carbon dioxide, a major contributor to global heating by the greenhouse effect. So cutting down the rain forest doubly increases CO_2 levels by the loss of trees and by their burning. British tropical biologist Norman Myers speculates that with the greenhouse effect, "Vegetation zones start to migrate away from the equator with all manner of disruptive repercussions for natural environments . . . . In effect, we are conducting an irreversible experiment on a global scale with Earth's stock of species . . . Certain evolutionary processes will be suspended or even terminated." As Harvard's E. O. Wilson once wrote, "Death is one thing; an end to birth is something else."

Just as unique as the flora and fauna of the Amazon are its aboriginal tribal people. Their rich and diverse cultures contain an irreplaceable knowledge base of the forests. The Brazilian constitution guarantees native people the land they occupy and exclusive use of all of its natural resources, but guarantees are only good if they're respected. Over half of the tribes haven't even had their territory recognized or delineated, yet development and occupation are already taking place in their lands. Even those with reserve boundaries constantly lose land by expropriation for mining, logging, agriculture and hydroelectricity. It's a tragically familiar story of aboriginal people around the world.

The preservation of the Brazilian rain forests depends on the successful struggle by indigenous people to control their land. The rain forest embodies their history, culture and identity — it's their only home and they will not destroy it. We in the industrialized countries can help by supporting them.

Saddled by its crushing international debt, Brazil looks to megaprojects that exploit its vast natural resources as the solution to the debt. Industrialized countries — Japan, those in the European Common Market, the United States — eager to secure resources in an increasingly depleted planet offer the dollars. Those resources are reaped with shockingly little regard for the long-term consequences on either the aboriginal people or the forest.

Brazilians perceive Amazonia as a cornucopia of resources waiting to be harvested. Trees like mahogany fetch a handsome price on the international market, and so do high-grade iron ore, copper, manganese, nickel, bauxite and gold. There are rumours of large fields of oil and gas somewhere in the Amazon. The vast network of waterways feeding the Amazon holds promise of giant hydroelectric projects. The forest is regarded as an impediment to economic development of the region. Clearing the trees yields huge tracts of land for farming, ranching and speculation.

Because the forest blocks "progress," the government's priority is to clear it. Timber is logged and hauled out without regard to replanting or destruction of other trees and plants. A network of railway tracks and roads pushed through the forest gives access to land that is cleared by cutting and burning. If you only travelled through Brazil by rail or road, you'd never know there was a tropical forest in the country! But the cleared land can't support farmers for more than a year or two and then the peasants abandon it to clear more forest.

The real wealth of the region rests in cattle. Heavily subsidized by government, cattle ranchers own huge blocks of land — though only 0.7 per cent of all landowners in the Carajas region own property over one thousand hectares, that small group occupies fifty-one per cent of the available farmland! But even ranching is not sustainable. Much of the secondary growth is not edible for cattle and some is even toxic. The scrub is burned repeatedly, each time destroying the soil and inviting invasion by hardier weed species and pests. So ranchers clear more and more forest for pasture that will last for only a few years. And the vast biodiversity of the rain forest is replaced with a predominance of two mammals — cattle and people.

When cleared of flora and fauna by cutting and burning, soil that is normally shaded under the canopy of the forest is baked hard by the sun. The carrying capacity for water is reduced, the amount of rainfall drops and winds roar through the cleared areas. So the effects of clearing ripple far beyond the plots, affecting the atmosphere, local weather and perhaps global climate in ways we don't understand.

In Brazil, as everywhere in the world, global economics underlies development. We in the industrialized countries, by our profiteering and consumer demands, compel countries like Brazil to convert their irreplaceable and priceless biological wealth into cash, often merely to service interest on debts. We must take action to help to preserve the forests.

There are groups fighting to staunch the terrible destruction: Probe International, World Wildlife Fund, Friends of the Rainforest and Survival International.

Canadians who care about tropical rain forests should make sure their money does not support destructive projects in the Amazon. Canadian agencies like CIDA should be urged to support sustainable activities that do not endanger the rain forests or the indigenous peoples. But it is organizations like the World Bank, the International Monetary Fund and multinational corporations that fund the most destructive projects. Finance Minister Michael Wilson casts the Canadian vote at the World Bank and has stated that environmental protection is a prerequisite for Canadian support of loans. The World Bank is currently considering a second Private Sector Loan of $500 million to Brazil to begin a devastating series of dams on the Xingu River. We should urge Mr. Wilson to establish Canada's credibility on environmental issues by announcing he will oppose the loan.

All six major Canadian banks are contributors to a commercial bank loan of $5.2 billion to Brazil. We should all write to our banks to determine what the social and ecological consequences of that loan will be. Amazon rain forests provide us with the opportunity to confront our involvement in ecological destruction around the world.

Wilderness to the Axe

*E*ven in our own country, the effect of our extractive mentality is increasingly obvious as wilderness falls before the brute power of a technology applied to fulfill the demands of an exploding human population with an insatiable appetite. Curves of resource depletion, habitat destruction and species extinction indicate that we are the last generation that will be able to make choices about wilderness — in fewer than thirty years, there will be none left.

British Columbia is blessed with a disproportionate share of the planet's wilderness. And it is here that the divergent views on the value of wilderness become clear. In B.C., logging interests carry tremendous clout, and there is a stiffening determination not to let native land claims and environmental concerns impede the continued destruction of its great forests. The prohibition of logging in protected wilderness areas, it is argued, represents an enormous "waste" and an unacceptable economic cost.

How much land do other parts of the world set aside? Development-oriented Alberta saves 9.1 per cent while U.S. logging states, Washington and Idaho, protect 11.3 and 9.7 per cent respectively. Crowded California protects ten per cent while Alaska sets aside a staggering 37.9 per cent of its land base. Even an impoverished Third World country like Tanzania saves twenty-five per cent of its land. New Zealand, the mother country of the major shareholder of B.C. Forest Products, the company poised to log B.C.'s Stein Valley, shelters 17.1 per cent of its forests. So how does B.C. stack up? The province protects a paltry 5.3 per cent — almost *ninety-five per cent* is up for grabs! For a province so dependent on its environment for tourism, recreation, fishing and forestry, B.C.'s record of wilderness preservation is shocking. How can anyone claim that more proposed wilderness parks or reserves are unacceptable infringements on logging? As a British Columbian, I am ashamed at what we are doing with our heritage.

Indications are that short-term profit priorities in B.C. will allow the continued destruction of what could be maintained for future generations in perpetuity. Yet all the areas in contention (the Stikine River, Canada's Grand Canyon; the Khutzeymateen, a grizzly-bear stronghold; Meares Island; the Stein Valley and others) add up to less than two per cent of B.C.'s land base. The challenge is not to maintain employment by cutting down the dwindling virgin areas, but to create jobs processing trees after they are down. No raw logs should be allowed to leave the province or country.

British Columbia lacks a coherent policy on preservation of wilderness. Its "management" programs seem to be designed to appease interest groups rather than based on real information. There hasn't even been an inventory of existing forests — we don't know what we have and will have in the coming years. Politicians are making up policy as they go. Activities with enormous environmental consequences are being proposed or carried out in B.C. without any historical and philosophical perspective. And blanketing all this is the refusal of the province to even consider native land claims. It's a surefire recipe for chaos and confrontation.

The dilemma is starkly clear in the Stein River Valley. This spectacularly beautiful virgin forest of 105,000 hectares is the largest wilderness watershed left in southwest B.C. It is sacred territory for the Lytton and Mount Currie Indian Bands who have lived in the area for thousands of years. The most recent government committee on wilderness recognized the importance of the Stein to native people by recommending that no development be allowed in the watershed without their approval. At least four different studies have concluded that it is not economical to log the Stein unless there is an enormous increase in the price of wood. Thousands of environmentalists and natives have already indicated their support for preservation of the Stein.

Nevertheless, Minister of Forests Dave Parker, after hearing submissions and representations from the Share the Stein Constituency Group, chose to ignore land claims, environmental concerns and economic analyses. In October 1987, he gave the go-ahead to log the Stein. It is claimed that *only* 9,000 hectares will be logged so the valley will still be preserved. But wilderness is by definition *untouched*. And what will happen to those jobs that depend on logging when the 9,000 hectares are gone? The same arguments about the sanctity of jobs and the economy will be dredged up again to justify still more cutting. Those trees in virgin forests cannot be reproduced and their likes will never be seen again. And once the watershed is invaded and the trees down, there is no issue. Logging the Stein only postpones some hard decisions that will have to be faced.

Across Canada we need a well-thought-out policy on the environment. Decisions can no longer be made on an ad hoc and issue-by-issue basis. Ecological studies indicate that even where pockets of wilderness are protected, extinction may still continue because truly balanced ecosystems are dynamic and require constant exchange of migrant organisms. We have to evolve a "network" of interconnected wilderness oases that will give us a chance of preserving a bit of the glory that once was.

There are also very selfish reasons to consider the legitimacy of native land and spiritual claims because their relationship with the land is radically different and makes ecological sense. Skirmishes over local issues preoccupy and deflect us from the long-term question: after all the wilderness is gone, what will the children of loggers and their children have to work with?

Trouble in the Forest

*I*n the populous part of Southern Ontario, farms and cities have long displaced the vast forests that once existed. So environmental change is not as obvious as in the clearcut mountains of British Columbia.

As a teenager, I read a short story in *Reader's Digest*. It told of a farmer who went to New York City to visit a friend. In the heart of the city, the farmer suddenly stopped and exclaimed, "I hear a cricket!" His friend scoffed that amidst all the din of traffic and people, he couldn't be hearing a cricket. The farmer responded by taking a coin out of his pocket and flipping it into the air. When it landed on the concrete, several people immediately looked around for the coin. The visiting farmer had made his point: you notice what you've been conditioned to notice.

When my wife was pregnant, I was amazed at the population explosion — there seemed to be expectant women everywhere. While the baby was nursing, I noticed women breastfeeding their infants all over the place. Remarkably, now that my children are older, the population boom has abated and women have stopped nursing their babies in public. We blithely go about our lives, unconscious of much of what's happening around us. It's only when we have become sensitized to something — like a cricket's call or pregnancy — that we take notice.

In October 1988, I saw a film that opened my eyes to what was once completely invisible but is now glaringly obvious. Filmmakers Holly Dressel and Gary Toole live in the rural part of eastern Quebec. A few years ago, they became aware that something was terribly wrong in the woodlot of their farm — the trees seemed to be sick. Investigating, they discovered to their horror that the deciduous forests in Quebec and much of eastern North America are dying from the effects of the increase in soil acidity from acid precipitation. Their story, *Trouble in the Forest*, documents the dying stages of Quebec's famous maple sugar trees. All indications are that the terrible blight is spreading across the northeastern part of the continent and affecting more than the maple trees.

It's been going on for years. Why haven't we noticed? Part of the reason is that we take plants for granted and don't notice them. They aren't cute and cuddly like animals, nor do they move or make sounds. They just stand there, silently storing the energy of sunlight in molecules of sugar, cleansing the air and clinging to water and releasing it into the air. If trees could talk, their collective screams around the world would be deafening.

Even though scientists in Scandinavia and Germany have known for years that forests were being adversely affected by acidity, there has been a great deal of denial in North America. Former U.S. president Reagan refused to acknowledge the reality of acid rain as an important ecological factor, even suggesting it might be a by-product of the trees themselves. Meanwhile, once you know the signs of trees dying from acidity, you can see them everywhere. The leaves are often discoloured from nutritional imbalances and the tree seems less bushy because there aren't as many leaves. In late August or early September, trees may change colour and the leaves fall in a premature autumn syndrome. The most obvious characteristic of "dieback" is the topmost branches — they die first and protrude like naked bones above the leafy part of the tree. Then over the next five or six years, more and more branches die while the tree rots, so its wood can't even be used for burning.

The billion-dollar maple sugar industry is an integral part of Quebec's economy and cultural history, yet the government has admitted that within a decade, the province's sugarbushes will be gone and the deciduous forests could be dead in twenty years! Driving along Toronto's Don Valley Expressway through the heart of the city, or along Highway 401 in early September, we can no longer avoid the obvious — what has afflicted Quebec is happening in Ontario too.

The ecological consequences of deciduous dieback are catastrophic. The flora and fauna of the forests are changing completely. It's already happening — birds depending on heavy foliage are being driven from the forest, and others like the pileated woodpecker that exploit insects in rotting wood are having a field day.

As the forest changes, wild animals don't just move off to a healthy part of the forest and wait until they can return. The forests are already fully inhabited, and when animals are driven from an area by dieback or clearcutting they have to compete for space in someone else's turf. Most don't survive. Much of the effect of acidity on the trees is due to the changes in the soil. But soil isn't just dirt, it's a living system, a vast population of microorganisms too numerous to count. If trees and the soil are being changed, do we dare think that we are not also part of that biosphere being affected as well? There will be an effort to find more acid-resistant trees and reforest the land with them. And no doubt something green will grow where the deciduous trees once flourished. But we will have lost the original community of organisms forever. And, just as a canary's death once warned coal miners that the air was foul, the Quebec maples are the latest canaries, an urgent warning we can no longer afford to ignore.

The Really Real

*I*n order to change public awareness and attitudes, we will have to inform and educate, especially through the media. But if individuals must take responsibility for judging the information that assaults us, we must understand the nature of information itself.

Television has great power to bring evocative images in vivid colour and motion. They are a major part of our experience of the world today. Our reality is based on experience that we sense through our nervous systems. Nothing is more convincing than when we see something with our own eyes. It's hard to believe that those impressions are incomplete.

But reflect on this: our eyes can only detect electromagnetic radiation within a very limited range of visible wavelengths. Ultraviolet and infrared, which lie on either end of our detectable range, are invisible to us, yet quite obvious to bees or rattlesnakes respectively. Similarly, a dog experiences a very different world from ours through its nose and can sense a reality that to us is unnoticeable.

The same limitations apply to our senses of taste, smell, hearing, touch. And not only do we detect just a fraction of all potential information available, but we alter a lot of the input we do get through the filters of our own preconceptions and experiences. Thus, a Stone Age person who witnesses a television program for the first time, though he or she would share the same visual and auditory impressions with us, would nevertheless see and hear something very different. It's not surprising, then, that it is often difficult to get two people present at the same event to recount it exactly the same way. Reality as we describe it is incomplete and a subjective construct of the human brain.

Scientists know reality in another way. It is based on a different mode of knowing and one that frequently flies in the face of our senses. To a scientist, an atom, a black hole or the fifth dimension can be every bit as real and tangible as a rock. Laws of probability are real as well: we know that a healthy octogenarian who has smoked two packs of cigarettes a day since his teen years doesn't disprove the relationship between lung disease and smoking. Many find this difficult to grasp. There can be a blurring in the distinction between the statistical and the anecdotal because individual stories seem far more real and concrete.

I saw this blurring happen when I was on a committee on science education for the Science Council of Canada. One part of the study was a detailed profile of science teachers and science course offerings across Canada. This involved the accumulation of a massive amount of information and its analysis and presentation in graphs and tables. At the same time, case studies were carried out. They involved researchers who went and visited randomly selected schools and observed the nature of teacher-student interactions in classrooms. What was striking was that the case studies had an impact that far outweighed the statistical analyses. The case studies were rich in anecdotes and real human situations, whereas the other involved impersonal, dry numbers that nevertheless gave a more accurate profile of what the situation was.

The same thing happened in a one-hour television broadcast several years ago on the CBC. It was a look at the current state of cancer treatment and considered the question of whether the "cancer establishment" had a vested interest in the puzzle that is cancer *not* being solved. Part of the program concerned the banning of the controversial drug laetrile in Canada. Scientist after scientist was interviewed and offered overwhelming evidence that laetrile was completely ineffective.

Then at the end, the Hollywood actor Red Buttons described his wife's battle with terminal cancer. She had received chemo- and radiation therapy and doctors had concluded that there was no hope. Buttons went on to say that he took his wife to Germany where she was treated with laetrile and was now considered cured.

Now any scientist knows that such a report does not *prove* a thing — remissions may happen for all kinds of reasons, but one case does not allow us to pick out the factor responsible. Yet the power of the anecdote was irresistible — with that one story, all the scientific evidence seemed insignificant. Cancer experts were furious and I don't blame them.

You can see the power of anecdotal reports on news shows when reporters discuss an issue and then turn to person-on-the-street comments. (I've done a lot of those "streeters" and there is nothing significant in them as a sample of opinion.) If the report includes two very strong pro comments and one weak anti position, the whole story is completely slanted. Yet those interviews are nonrandom samples that are further selected by the producers of the show to have an impact.

Although we have to distinguish between anecdotal material and quantifiable scientific data, we must never forget that science itself is an activity carried out by human beings who have all the perceptual baggage of their society and personal experiences.

Ever since the ancient Greeks posed the question What is reality, philosophers and scholars have pondered through the ages. Far be it from me to tackle the question, but I can make some observations as a scientist and journalist.

Our enormous and complex brain gave us behavioural flexibility; we can think in abstractions, innovate, synthesize, project consequences of present actions in the future, learn, remember and share knowledge. Other species may have some of these capabilities but the output of our brains is unique on this planet. We are no longer bound by the restrictions of our neural circuitry; indeed, we have created machines that are incredibly faster, more reliable and efficient to add to our brains.

Through the 50,000 to 100,000 years of our species' existence, we have scarcely changed anatomically or genetically — Cro-Magnon man would not be noticed on the streets of Toronto if he were dressed in modern garb. Yet in the millennia since *Homo sapiens* appeared, that brain has generated the enormous outpouring of language, song, poetry, culture and civilizations around the planet!

But if the sensory apparatus of all people is essentially the same, do we not begin with a common base of sensation of the world around us? After all, light waves focussed on the retina elicit the same electro-chemical impulses down neurons of all people. Aren't time, space, sound, experienced through the same nervous system? And here is where this remarkable organ begins to surprise us.

If you have ever experienced a sudden life-threatening situation or spent an agonizing wait for a special occasion, then you know that time expands or contracts depending on our "state of mind." Try going to a movie or watching a television program with someone and then discuss it later — it's amazing to compare notes and observe the differences in what is memorable or significant.

In my opinion, one of the most profound films ever made is the Kurosawa classic, *Rashomon*. In it, a woman and her husband are captured by a bandit who rapes the wife, then kills the man and escapes. When he is caught and brought to trial, the film flashes back to the events through the eyes of the woman, the bandit, the ghost of the dead husband and a man who chanced to see it from the bushes. Each story is completely different!

Each of us, however closely we may share a culture and background, is a unique combination of genes and experiences that shape the perceptive capacity of our brains. We are constantly filtering out our experiences and creating a reality that is in essence an edited version of events.

In the film *The Big Chill*, there is a memorable scene in which one of the characters is accused of a rationalization.

"Of course," he answers, "rationalization is the most important thing in life."

"More important than sex?" he is asked, to which he replies, "Of course. Have you ever gone for three weeks without a rationalization?"

I agree. We are the great rationalizers — after a violent argument, both parties to it will reconstruct the events to cast themselves in the best light. My point is that all people feel that their view of the world, which has been shaped by their culture and experiences, is real. But face someone from India, Botswana and England with a similar set of circumstances such as a serious illness, birth, marriage or death and you will witness reactions that are truly worlds apart.

Scientists regard science as a unique way of learning about the world because it provides a means to reproduce events in time and space, and thus provides a picture of objective reality. But the history of science puts the lie to that. Scientists are first and foremost human beings, as subject to the same limitations of cultural bias as anyone else.

So it should not be surprising that Charles Darwin, a member of the British aristocracy, educated in a period of slavery, colonialism and mercantile expansion, should have described his great insight — evolution by natural selection — in terms of struggle and selection of the fittest. Today, cooperation and sharing are also seen as components of evolution. Geneticists at the turn of this century used their science to provide a biological basis for the "inferiority" of blacks, the poor, homosexuals, criminals, gypsies and many others. Today, geneticists realize their science cannot make such evaluations. Neuroanatomists "proved" that white males had larger intellectual capacity than white females who had a higher capacity than blacks and native Indians, notions that are rejected now.

The list is long. This is not to say that scientists are bigoted or evil, only that they cannot escape their own as well as their society's preconceptions and social values. Nor is there any indication that scientists are any more enlightened or objective today than they were in times past.

Unfortunately, too few scientists or members of the general public are aware that data are not just data, that cultural framework, values and assumptions affect the kinds of problems that are perceived, the way experiments are carried out and the interpretations of the results.

Even when we understand the pitfalls of scientific "objectivity," we are still stuck with the perceptual limitations imposed by our senses. This may no longer be enough to survive the hazards of a radically changed world of the last decade of the century.

Two grapes tainted with cyanide were enough to bring a multi-million-dollar Chilean agribusiness to its knees. The discovery of those two fruits must represent an incredible achievement by the customs office. On the basis of an anonymous phone tip and those two poisoned fruits, several governments, including those of Canada, the United

States and Japan, immediately imposed a ban on all Chilean fruit and vegetables. Yet those same governments maintain a remarkable tolerance to wide public exposure to the chronic presence of PCBs and potential hazards like Alar in North American–grown food. We are able to respond heroically to an immediate perceived crisis far more readily than we can to much more significant global problems. Why?

That is the question posed by two Stanford University professors, psychologist Robert Ornstein and ecologist Paul Ehrlich, in a provocative book, *New World, New Mind* (Doubleday). They remind us of numerous contradictions in human behaviour. We can be deeply moved by reports of a child trapped in a well in Texas while ignoring the deaths of twenty million infants a year from diseases that are preventable. We are mesmerized by images of three whales trapped in ice off Alaska but seem unconcerned about global *extinction* that is now estimated to claim more than *two species an hour*. Threats by a small band of terrorists create widespread fear, yet we tolerate the carnage on North American highways that claims over 50,000 lives a year and the continued enlargement of an already massive global arsenal of nuclear weapons.

For Ornstein and Ehrlich, the explanation of these inconsistencies lies in our long evolutionary history and the speed with which changes are now occurring. We detect events happening in our surroundings through the apparatus of our senses. Our reality is delineated by the range of those senses. But in spite of such limitations to our sensory input, we are able to respond magnificently to our environment. Ornstein and Ehrlich point out if we are in a cave and the shadow of a bear crosses the entrance or if we are standing on a branch and we hear it crack, we are able to respond immediately and take protective action.

As well as coordinating the input from our senses and sending out appropriate responses, the human brain also invented an idea called a *future*. And because we recognize that what we do today can affect what happens in the future, our species is uniquely able to select from a number of options to maximize survival. We have deliberately chosen a path (albeit strewn with dead ends and wrong turns) into the future, and it has worked so well that we are now the most ubiquitous and numerous large mammal on the planet.

But our enormous increase in numbers coupled with an acceleration in technological muscle power conferred by science has created a "new world" in which our "old minds" are no longer adequate. For most of human history, small groups of hunter-gatherers were not even aware of the existence of others of their kind in distant places. Their world was circumscribed by the physical constraints of communication, travel and food. When human numbers were small and technology simple, their environment could easily absorb and recover from the impact of human activity. Human beings only reached a billion early in the past century, yet we have already passed five billion and appear on the way to doubling in fewer than another fifty years. While our numbers are now vast, we continue to respond to immediate and personal crises rather than to those that affect our species as a whole. Where once the hazards to be avoided were immediate and tangible, today they are more abstract and lie years away.

Many dangers we face are now beyond the detection capacity of our sense organs. We can't tell that there are pesticides in our food, dioxin or PCBs in water, or an increase in low-level background radiation or radioisotopes. We are unaware of the ozone layer or changes in it or an increase in CFCs or CO_2 in the atmosphere. Acid rain doesn't register on us physically. And since we have adapted to respond to threats that we can "feel," we find it difficult to take the perils of the "new world" seriously.

Ornstein and Ehrlich believe that the constraints of human perceptions that affect our priorities and actions are not immutable consequences of our genes. Instead, they feel that it is possible to develop a "new mind" through a better understanding of the global community of organisms of which we are a part and of the finite resources that must be shared and recycled by all forms of life. The massive response of governments to two poisoned grapes should be an occasion to reflect on what we must do to counter the *real* hazards of the new world.

Rx FOR A SICK PLANET

Introduction

*T*he environmental crisis is no longer in dispute, but what can be done about it? In some places, the long-term aims of environmentalists clash with short-term demands for jobs and profit. But who is right?

Severn, my ten-year-old daughter, has been an activist all her life. She was eight months old when she went on her first peace march and wrote her first letter to the prime minister when she was five. She routinely sells old books, toys, cookies and lemonade to raise money to save wilderness areas. One day when she was eight, she looked at me and asked, "Daddy, we think we're right in fighting to save these forests, aren't we?" I vigorously nodded my head. "But the people who want to cut the trees down also believe *they're* right too," she continued. Then she added the zinger, "So how do we know *we're* right?"

Children have a remarkable ability to see with a clarity unclouded by history, vested interest or belief system that so often blind grown-ups. While Severn already supports peace, the environment and native people as her parents do, her indoctrination is far from complete. Every adult would do well to consider her question, "How do we know we're right?"

My answer to Severn was that we don't know for sure, but if we succeed in saving some forests and we turn out to be wrong, all that will have been lost is some immediate profit — the trees can always be cut later. But if the logging industry cuts all the trees down and *they* are wrong, those forests can never be replaced.

In order to find new solutions to the crisis we face, we have to get rid of many of our outdated assumptions, see through the smoke screens generated by power groups concerned with short-term interests and seek profound answers.

Red Herrings Galore

*I*n the debate over disappearing wilderness areas and what should be done, politicians often reflect the preoccupation with profit and development. Even environment ministers seem more concerned with how to justify development than with protecting unique areas.

A cabinet minister's response to a query during Question Period in Parliament is usually full of rhetorical mockery, indignation or accusations but rarely includes an answer. It's often the same in environmental disputes — red herrings are continually thrown out in response to points made by environmentalists and native people, thereby deflecting attention from the fundamental issues being raised.

In a letter to *The Globe and Mail* (Dictating to the North — June 10, 1988) Tobin Lambie of North Bay defended trapping, mining and mineral exploitation as "traditional northern pursuits."

About the controversy over the fate of the Temagami wilderness in Northern Ontario, he goes on: "It's a shame a small group of Southern Ontario environmentalists looking for a playground can dictate what happens in the North. The southern bleeding hearts should worry about their own backyard, not ours."

It's a classic example of the way serious environmental issues are avoided by defenders of continued exploitation of wilderness. B.C. Forests Minister David Parker constantly blames "outsiders" for poking their noses where they don't belong. The South African government uses the same rationale to tell the rest of the world to mind its own business on apartheid. But the fate of wilderness is a matter of concern to *everyone* for economic, health and spiritual reasons.

Today, biologists consider the rapid loss of biodiversity, i.e., species and community extinction, to be a global catastrophe. The apparent determination of Canada's logging industry to get at every remaining untouched watershed to maximize profits has to be viewed in the global context of a frightening loss of biological variety.

No one should apologize for supporting the preservation of the Stein

Valley in British Columbia or Temagami forest in Ontario, or for that matter, forests in Tasmania, Sarawak or Brazil, even though he or she doesn't live nearby — the wilderness is vital to all.

"Multiple use" is the current buzzword to cover clearcut logging and other ways of "developing" a wilderness. This is the forest industry's favoured strategy for exploiting the economic potential of the Stein. Supporters of native people's demands to preserve the watershed are accused of threatening the forest industry. Seen from the perspective of the mills in the area, a case might perhaps be made for the local, short-term benefits of cutting the trees in the Stein. But as the millhands in Hazelton, B.C., recently learned, the forestry industries are not "sustaining the yield," so when trees run out, the mills are shut down. The forest industry, which claims that forests are forever, is yearly cutting down old-growth trees amounting to twenty times the total found in the Stein. Removal of this one watershed from annual cuts would have an imperceptible effect on the torrent of logs pouring out of B.C. forests.

Other ways are used to obfuscate issues. In Lytton, B.C., in April 1988, Chris O'Connor, the most visible forester supporting logging of the Stein Valley, tracked me down. He is an employee of Lytton Lumber Ltd., a mill with much to gain from logging the Stein, and the son-in-law of owner Lew McArthur.

O'Connor claimed to have seen "all kinds of" Mercedes and BMW cars belonging to people who attended last year's Stein Festival, an annual watershed celebration. My initial reaction was disbelief, but my second was, so what? What's the point? I assume O'Connor is suggesting the owners of such cars are well-off. If that's true, then we should rejoice that some of the wealthy are looking beyond making money to support wilderness preservation. But in fact what do a car, clothes or appearance have to do with the fate of the Stein? Nothing.

Former Alberta Environment Minister Kenneth Kowalski publicly described environmentalists as poorly dressed drug consumers who have been mentally damaged by high altitudes. In this revelation of his own ignorance and bias, Kowalski dismissed literally thousands of older people, middle-class professionals, parents and youngsters who are involved in all kinds of environmental issues.

Ironically, Kowalski was supposed to be a protector of the environment, yet he acted as if environmentalists were his enemies. Even if environmentalists were as the minister describes them, what relevance would it have to the issues? None.

B.C.'s Parker has suggested that *tenure* somehow disqualifies academics from debate. At a talk I gave in the spring of 1988, Les Reed, a University of British Columbia forestry professor and outspoken supporter of logging both the Stein and South Moresby in the Queen Charlotte Islands, boasted that unlike me, he lacks tenure.

It was an astonishing admission, because tenure allows academics to speak openly without fear of political reprisal. In contrast, Reed depends on forest industry contributions for his salary and research. These monies are then matched by the federal government. So if industry doesn't contribute, his research support and salary are gone. How free is Reed to be critical of his source of support? Tenure avoids such a debt to a vested interest.

Let's not waste more time with red herrings and get to the issues.

The Power of Diversity

*I*n the past decades, the scientific community has undergone a tremendous expansion while knowledge has increased proportionately. However, too often the accumulation of information is mistaken for knowledge that provides understanding and control. We can't afford to make such an assumption because it fosters the terrible illusion that we can "manage" wilderness and has resulted in destructive consequences.

Globally, old-growth forests are being cleared with alarming speed. In the past two decades, geneticists have made a surprising finding that foresters should heed. When seemingly homogeneous populations of organisms were analysed using molecular techniques, they were unexpectedly found to be highly diverse. When looked at from individual to individual, the products of a single gene are found to vary considerably. Geneticists call such variability *genetic polymorphism*, and we now know that a characteristic of wild populations of any species is a high degree of genetic polymorphism. Apparently, maximum genetic diversity optimizes the chances that a species can withstand changes in the environment.

When individuals of the same species are compared, their patterns of genetic polymorphism differ from region to region. Thus, whether a tree, fish or bird, different geographic subgroups exhibit different spectra of variation. So Ronald Reagan was dead wrong — if you see one redwood tree, you *haven't* seen them all. Stanford ecologist Paul Ehrlich says, "The loss of genetically distinct populations *within* species is, at the moment, at least as important a problem as the loss of entire species."

This biological value of diversity can also be applied to a collection of species as well. A forest is more than an assemblage of trees, it is a community of plants, animals and soil microorganisms that have co-evolved together. This aggregate of species creates a highly resilient forest with a great capacity to recover from fire, flooding, landslides, disease, selective logging or storm blowdowns. That's because the diverse species remaining in the surrounding areas can replenish the damaged parts. Clearly, we should try to maximize forest diversity by protecting as many different old-growth forests as possible. That's the best way to ensure the maintenance of a broad genetic base on which the future of the forestry industry will depend.

There is a way to illustrate the power of diversity by looking at our own bodies. Just as a forest is made up of vast numbers of individuals of different species, *we* are an aggregate of some 100 *trillion* cells that vary in size, shape and function. These different cells are organized at many levels into tissues and organs that all come together in a single integrated whole — a functioning body.

The collective entity that is each of us thus is a mosaic of an immense array of different cell, tissue and organ types that have enormous resilience and recuperative powers. If we suffer a cut, bruise or infection, the body has built-in mechanisms to overcome the assault. We even have the ability to regenerate skin, liver, blood and other body parts and compensate for damage to the brain and circulatory systems. We can function pretty well with the loss of some body parts such as a digit, tonsils or teeth. In short, our bodies can absorb considerable trauma and recover well, a tribute to the cellular diversity in form and function.

If we amputate large parts of the body, we can still function and survive. Thus, we can live with the loss of limbs, eyes, ears and other parts, but now each loss confers greater dependence on other people and human technological ingenuity to compensate for lost abilities. With the power of modern science and high technology, we can make artificial substitutes for teeth, bones, skin and blood and we have even devised machines to take over for the heart, lungs and kidney. In principle, it should be possible for an individual to survive the combined loss of organs not absolutely necessary for life and those that can be mimicked by machines. Thus, a blind, deaf, quadruple amputee who is hooked to a heart-lung and kidney machine could live and would still be a person, but one with capabilities and resilience radically restricted in comparison with a whole individual. Essentially, such a patient would be a different kind of human being, created by and dependent upon human expertise and technology.

In the same way, a forest bereft of its vast biodiversity and replaced by a limited number of selected species is nothing like the original community. It is an *artifact* created by human beings who foster a grotesque concept of what a forest is. We know very little about the basic biology of a forest community, yet roadbuilding, clearcut logging, slash burning, pesticide and herbicide spraying, even artificial fertilization have become parts of silviculture practice. The integrity of the diverse community of species is totally altered by such practices with unexpected consequences — loss of topsoil, death from acidification, weed overgrowth, disease outbreak, insect infestation and so on. But now, caught up in the mistaken notion that we have enough knowledge about forests to "manage" them in perpetuity, we end up ricocheting from one contrived Band-Aid solution after another. Medical doctors today are struggling to readjust their perspective to treat a patient as a whole individual rather than as an aggregate of autonomous organ systems. A similar perspective has to be gained on forests. The key to development of sustainable forests must reside in the maintenance of maximum genetic diversity both within a species and between the species within an ecosystem. If we begin from this basic assumption, then the current outlook and practices in forestry and logging have to be radically overhauled.

Owning Up to Our Ignorance

*C*urrent practices in forestry are encouraged by the use of language. On a phone-in radio show about the controversy over whether to log the Stein Valley in B.C., Pat Armstrong, the paid spokesperson for the logging interests, defended "multiple use" of the watershed. (Incidentally, all but one participant on the hour-long program opposed logging the valley.) When one caller referred to the "devastation" and "destruction" caused by clearcut logging, Armstrong bristled and said he refused to enter a discussion involving language like that.

Armstrong had correctly recognized the power of language. The words we use reflect and shape our attitudes and values. But the forestry industry that Armstrong defends so vigorously is itself a perpetrator of illusions created by the use of words. Foresters refer to "decadent" forests, meaning the stage in the evolution of a forest at which a lot of economically valuable trees are approaching the end of their lives and will die. The implication is that they should be "harvested" to avoid a "waste"; yet for millions of years, forests have matured and changed while supporting a multitude of organisms.

Our Forest Services apply herbicides to forests in order to kill "weed" species of trees. Again, the word suggests a type of tree that is a pest or has no use. We hear of "thinning" or "culling" trees and the cutblocks are referred to as "tree farms." Foresters talk about "plantations," "crops," "standard forestry practices" and "managing" wilderness areas. All these words have their roots in agriculture and carry the implication that forestry is simply the farming of trees.

We should not forget that agriculture is a sophisticated activity that is over ten thousand years old. Even though it has a long history, we have instituted such devastating practices as the heavy application of chemical pesticides and herbicides (thereby poisoning workers, consumers, water and the soil), planting of vast acreages of single crops of uniform genetic makeup (which are extremely vulnerable to disease and create dependence on artificial fertilizers and other chemicals) and exposing much of the topsoil to erosion. In contrast, modern forestry is in its infancy and has never received the kind of financial support to maintain a top-notch, productive scientific community. The "crops" that foresters manipulate are not domesticated plants, and they must grow in uncultivated areas for decades. So the words used to imply that forestry has a similar basis as agriculture perpetrate a delusion.

Constant repetition of words and ideas reinforces a belief in their validity until they are assumed to be factual. A letter (*Globe and Mail*, June 29, 1988) from Dave Parker, a forester, who is now B.C. minister of Forests, provides an illustration. He defends destructive clearcut logging as:

> an accepted practice which follows the principle used by nature: clearing areas to regenerate forests. The difference is that nature uses insects, disease and wildfire. Nature's technique of removing hazardous material, controlling insects and disease and preparing a site for new forests by fire is emulated in forest management through prescribed burning . . . .
>
> In British Columbia, we manage our forests on the basis of integrated resource management — considering all demands for use of the forest, whether those demands are for timber harvesting, recreation, preservation, and/or wildlife habitat, to name a few. In every forest management plan, all aspects of the ecological makeup of an area are considered.

A forest is a complex ecosystem made up of numerous organisms, many of which haven't yet been identified and whose behaviour and biological characteristics have not been defined. We have only the most superficial description of the inhabitants of a forest and almost no understanding of their interaction. How then can we consider "all aspects of the ecological makeup"? One thing that we have learned is that a major characteristic of wilderness areas is diversity — great variation in species composition, numbers of organisms and genetic makeup of each species. "Natural" forests do not grow back the way we reforest by planting uniform seedlings at regular intervals. Nor do we emulate nature when we apply pesticides, herbicides and fertilizers to the soil, clearcut huge tracts and destroy the waste by burning. Communities of organisms co-evolve over long periods and maintain a balance through natural selection. The immense machinery involved in clearcut logging destructively churns up the soil. The operation drives wildlife from an area.

We often forget that soil is a living community of organisms, not just dirt. Each forest floor has its own distinctive accumulation of organic material from countless generations of plants and animals that have lived in an area. In British Columbia, much of the logged area is on steep slopes where the topsoil is thin and there is much rain. Much of the soil is quickly lost before a new group of trees can take hold and cling to the mountainsides. It is the height of self-delusion to believe that clearcut logging imitates nature.

Of course trees, plants and animals can grow back where forests have been clearcut. After all, about ten thousand years ago, all of Canada was buried under an immense sheet of ice and there were no forests. But it took all those ten millennia to create the natural treasures that we now enjoy, and it is a conceit to think that we know enough to act so wantonly and then re-create forests as if they are tomato or wheat plants. To equate the areas regenerated after clearcutting with forests before logging is a gross devaluation of language. In the long-term interests of the forests that all of us want to share, we should stop misusing language to cover up our ignorance and inadequacies. This is not inconsequential: whole ecosystems are being destroyed under the impression that we can "manage" our resources.

At the beginning of August 1987, more than two thousand people hiked over a mountain pass to camp at the head of the Stein River Valley in British Columbia. They were there for a festival celebrating another of this planet's special places.

At the Stein Festival, I met a forester who had accused me on radio of being "too emotional" and having "little factual information" to back up my support for the preservation of the Stein. (I have never figured out what is wrong with being emotional about something that matters — I *am* emotional and I do not apologize for it.) But when I challenged his claim that foresters know enough to replace the likes of the trees they would cut down he replied, "We're almost there. We've learned a lot and we'll soon know everything we need." He was confident his children could look forward to logging the kind of trees he wanted to harvest in the Stein.

Scientists *have* learned a lot in the past half century. Many of their insights are truly mind-boggling — models of atoms and subatomic particles, black holes, DNA and the immune system. These investigations have been accompanied by the invention of technological tools to disrupt and alter much of the natural world. Our manipulations are often extremely powerful and yield immediate results, yet they are crude approximations of what exists in nature. It may be true, for example, that a mechanical heart is quite a technical achievement, but it would be a travesty to suggest that such a simple pump comes anywhere near duplicating the real thing. Unfortunately, people like that forester have equated our power to affect short-term changes in nature with long-term control and progress. This is sheer arrogance.

Remember, Canadian support of science has always lagged behind that of other industrialized countries, and forestry, not being a high-profile area, has always suffered a severe shortage of funds and top students. How, then, can we possibly believe we comprehend the diversity of organisms and the complexity of their interactions when our analytical techniques and expertise are so limited? Think about this.

I spent twenty-five years of my life practising science. My area of genetics has been a high-profile field that has attracted top scholars and funds for decades. My entire career was spent focussed on one of the estimated thirty million species on this planet — a common fruit fly. In fact, this fruit fly has been studied by geneticists since 1909, and for over seventy years, it has been at the centre in the study of heredity. Tens of billions have been invested to pay the salaries of thousands of geneticists who have devoted complete lifetimes studying the fruit fly. Four scientists who studied fruit flies have already won the Nobel Prize, and no doubt there will be more in the near future.

The results of all this effort have been impressive. It is possible to create all kinds of mutant strains affecting the fly's behaviour, anatomy and viability. We can take a single cell from a fly and "clone" it to create another fly. We can extract pieces of DNA, alter them and reinsert them into flies. We can make flies with wings growing out of their eyes, legs out of their mouths or antennae, four wings instead of two, twelve legs instead of six. We have gained tremendous insights and manipulative powers.

But you know something? After investing hundreds of thousands of person-years of research, billions of dollars of grant money and all the latest equipment in studying this one species, we still don't know how they survive the winter. We still can't understand how a fly's egg is transformed first into a larva then a pupa and adult, something every fly manages easily. There is a species of fruit fly so closely related to the one I've studied that only a handful of specialists in the world can tell them apart, but the flies have no problem. It turns out that we have very little understanding of the basic biology of this one species, and it is only one of tens of thousands of species of fruit flies! If we know so little after all that, how can we possibly think that we have accumulated enough knowledge to enable us to manage complex ecosystems like the Stein Valley? As a scientist, I have been overwhelmed not with the power of our insights but with *how little we know*.

We have become so intoxicated with our clever experiments and increasing knowledge that we forget to see the intricate interactions and nearly infinite complexity that exist in the living world. Instead, smug in our faith in the knowledge of scientists, we perpetrate the notion that we already know enough to cut down the last areas of wilderness. Those "scientists" and "experts" who speak so confidently about the logging industry's ability to mimic the virgin stands that are being cut down reveal that they literally cannot see the forest for the trees.

Stuck in Red Tape

As ecological problems become higher political priorities, it is glaringly apparent that the manner in which government departments are organized often ensures that we will not handle the issues in ways that make biological sense.

One of the earliest lessons I remember from school was the cycle of water — how it rains down from the clouds onto the soil, where it is absorbed by plants and transpired back into the air; or the moisture is soaked up by the earth to be released over time, into creeks, rivers, lakes and oceans from where it evaporates back up into the sky; or the water seeps deep into the earth to become part of the groundwater which may eventually well back to the surface. Even as a child, I could appreciate the complex interconnections in nature that are too often forgotten by adults.

The most powerful impressions one gains in studying biology is how interwoven and hence interdependent life is. However, society tends to separate natural phenomena into categories that can be dealt with politically in convenient bureaucratic subdivisions. And in the process we often lose sight of the whole.

In British Columbia, many of the once-great runs of salmon have been drastically reduced, and hundreds of streams and creeks no longer support runs at all. These losses can be traced to the destructive effects of human activity — urban development, logging, mining or introduction of large numbers of hatchery-raised fish that severely constrict the genetic base of the populations. So how do politicians go about trying to restore those great runs? An immediate problem is that the fish don't know about national boundaries. Thanks to Canadian taxpayers, salmon are reared and released in Canadian waters, yet they move according to their biological urges. So they can be intercepted by fishermen from the U.S., Japan, Korea and other countries far out at sea. Officials from different nations negotiate regulations on behalf of the vested interests of their countrymen, not the fish. As well, there is constant jockeying between federal and provincial ministries over control and territory. So right from the start, the "salmon problem" is chopped up among many levels of bureaucratic administration.

The adult salmon are sought by commercial fishermen, so obviously fishing falls within the domain of the department of Fisheries and Oceans. But *sport* fishing is a major revenue generator, so the ministry of Tourism and Recreation must also be involved. In order to handle the aboriginal rights to food fish, the department of Indian and Northern Affairs has to be brought in. A large part of our difficulty with managing salmon is sheer ignorance about their life cycle and behaviour, so the ministry of Industry, Trade and Technology must support basic research. Factors contributing to the salmon decline have been oil spills from tankers, logging, agricultural run off and urban sprawl, so the departments of Forestry, Agriculture, Transportation, and Energy, Mines and Resources have to be consulted.

We don't have mechanisms to deal with salmon in a biologically sensible way. By cutting the issue apart along jurisdictional lines, we ensure constant squabbles over political turf. We need an *ecological* approach that focusses on the fundamental issue — salmon enhancement — and subordinates all the other departmental constraints and infighting.

Another illustration of the hazards of bureaucratic compartmentalization is the Great Lakes. This massive body of fresh water is one of the world's great natural resources. The Great Lakes are an important waterway to transport goods and a major source of water for households, agriculture and industry. The lakes and connecting rivers are used as dumps for sewage and industrial waste. The Great Lakes fishery remains an important economic activity and the waters are also used to generate electricity. Of course, the shores of the Great Lakes are an esthetic and recreational delight, an inspiration to artists and poets. These have all been benefits for humans, but countless communities of fish, birds, mammals, insects and plants have depended on the Great Lakes far longer than we have. From the standpoint of human needs alone, management of the Great Lakes is a political nightmare. There are two federal governments involved, as well as two provinces and eight states. Then there are large cities like Chicago, Cleveland, Detroit, Hamilton, Toronto and Windsor as well as dozens of towns and villages that use the water for different purposes. So a number of ministries and departments at different levels of government have to be involved. But their jealous protection of political spheres, constituents and vested interests precludes commitment to a comprehensive program to protect the water for future generations. While negotiations move with glacial speed, unrelenting human activity continues to overwhelm the carrying capacity of the lakes.

We desperately need to incorporate an ecological perspective into our political apparatus, one that recognizes that the environment doesn't end at arbitrary human boundaries of cities, states and provinces or nations. Protection of the integrity of ecosystems ought to subsume all the other government priorities. There should be a Supreme Office of the Biosphere whose mandate is ensuring minimal ecological disruption from all human activity. Girdling that office will be Terrestrial, Aquatic and Atmospheric departments, all interlinked and the final authorities for ensuring that the mandate of the Biosphere Office is realized. All other traditional subdivisions of human activity — finance, industry, health, and so on — should fall under these three superdepartments. It may be a pipe dream, but it sure makes more sense than the way we're doing things now.

Borrowing from Children

*E*nvironmental awareness can be dramatically catalyzed by an issue that becomes a symbol. DDT, PCB and CFC now instantly summon up the "costs" associated with man-made chemicals. Bhopal, Chernobyl and Valdez illustrated the predictable occurrence of human error with all technologies. Whooping cranes, whales and baby seals raised the question of the innate value and rights of other life forms.

And increasingly, the fate of entire ecosystems — the Franklin River in Tasmania, Antarctica and the Amazon rain forest — have become contentious because they are the focus of opposing forces of preservation and development. In Canada, the Windy Bay watershed on Lyell Island in the southern part of the Queen Charlotte Islands off the northwestern coast of British Columbia focussed efforts by the forestry industry, native people and environmentalists for fifteen years. On March 21, 1987, I wrote the following column:

"We no longer inherit the Earth from our parents, we borrow it from our children." This is the stark message on a Green Party poster, a warning well worth some thought. Most of us hope that our children will be better off than we are, that they will have lives qualitatively richer than ours. But now it's not possible to cling to that expectation. Each day, the planet becomes poorer, as species of animals and plants disappear forever and the vast pool of biological diversity shrinks.

Even as recently as the early part of this century, it took two men several days to cut down one of British Columbia's immense trees. Today, one man and a chain saw can do it in minutes. But we haven't compensated for the environmental impact of our numbers and technological power by scaling down the way we attack it.

The first European settlers came to a continent teeming with life, but the way we exploited it is a terrible chapter in our species' history.

Passenger pigeons were once abundant beyond belief, literally darkening the skies for days on end when they migrated, yet we exterminated them within a century. The great herds of bison suffered a similar fate. How much poorer we are for their loss, to say nothing of the loss to the animals themselves.

Don't we learn from history? Do we not have any sense of our place in nature that should limit our predation? We have always taken pride in the fact that our brains enable us to see beauty in the world and even to add to it through poetry and art. Does it not diminish us when we cause the extinction of other life forms that we appreciate as unique and beautiful?

One of the most unusual ecosystems in all the world is British Columbia's Queen Charlotte Islands. Because some of the islands were not covered by ice during the last glaciation, plants and animals survived there that are found nowhere else in the world. And for thousands of years these islands, the forests and surrounding waters supported a rich and diverse civilization whose descendants are the Haida Indians.

Much of the northern part of the Charlottes has been denuded by logging, but the southern fifteen per cent, called the South Moresby Islands, is still relatively untouched. In this area is Lyell Island, which harbours a 2,800-hectare coastal rain forest called Windy Bay. It took thousands of years of plant succession and selection for the immense trees that now populate the watershed to grow. Some of them are more than one thousand years old!

For more than twelve years, environmentalists and the Haida have fought to protect the South Moresby region from logging by Western Forest Products Ltd., which holds the tree-farm licence in the area. Windy Bay has been the focus of the debate. During this entire contentious period, logging has been allowed to continue on the island around Windy Bay.

The British Columbia government's tactic has been to approve logging permits while its own public inquiry was going on. The first study group lasted three years and recommended the establishment of a more broadly based committee. So the South Moresby Resource Planning Team was set up, with representatives of logging, Haida, environmentalist and government groups. It deliberated for five more years and recommended the preservation of Windy Bay. This report was put on the shelf and ignored for two years, and all the while, logging continued.

Under public pressure, the B.C. government again set up a study group in 1985, this time called the Wilderness Advisory Committee, and charged it with looking at *sixteen* different disputed areas in the province. Furthermore, the government asked for recommendations on all of them within three months! Shortly after the committee's hurried report was turned in, a new government was elected and declared itself unbound by any previous studies. Logging has not stopped during the entire period.

The issue involved is not whether there should be logging anywhere — no one disputes the economic importance of logging. But the question is whether a few special wilderness areas ought to be left untouched. There are strong philosophical, esthetic and practical reasons for leaving some remnants of nature as it once was.

It is the height of hypocrisy and cynicism to allow logging to continue in the special areas while public debate about their fate is still going on. The tactic is transparent — once the trees are cut down, they are gone forever and there is no debate. Those who maintain that they can replace the likes of what we "harvest," that they can "manage" our resources and sustain the yield, cannot see beyond immediate profit and will not admit to enormous ignorance. Future generations will judge them harshly.

Only our present generation, through strong and sustained pressure on government leaders, can hope to preserve a remnant of what was once the great Canadian heritage.

* * *

Under intense pressure from the public, both the federal and provincial governments negotiated furiously. For British Columbia, the fate of South Moresby became the symbol of the future path for loggers, environmentalists and natives. I wrote this column on June 6, 1987:

By the time this column appears, the future of the South Moresby area of the Queen Charlotte Islands may have been decided. Federal and provincial negotiators have agreed on setting the area aside as a national park. The final formal agreement will be a major achievement by environmentalists, and all Canadians should congratulate Ottawa and British Columbia for the successful conclusion.

The fate of the area has come down to a matter of money. British Columbia's government says about $100 million should be set aside for the park, while the federal ministry of the Environment wants a spending limit near $40 million.

This is in the tradition of the political game of poker. In all the discussions, however, two issues have been omitted that cannot be ignored in future deliberations over wilderness areas: Are the only things of worth those that are measured in terms of their use for humans? Are the only meaningful values those that can be expressed in dollars and cents? As long as we assess everything as a potential resource, then it's just a matter of thinking up a use and costing it out to decide on the fate of anything.

In January 1987, I attended a meeting in New Zealand at which I heard a bioethicist justify research on human embryos because it would eventually "allow us to recover the many embryos that are normally aborted and so are an enormous waste." This "resourcist" attitude reminded me of a chemistry class in high school in which the teacher informed us that if a human body were reduced to its basic elements — calcium, carbon, sodium and others — those chemicals would be worth only a few dollars. (Perhaps on today's market a couple of hundred.) What impressed me at the time was the thought that those few bucks' worth of atoms when arranged as a living person had a value beyond measure. The teacher had given us a nonsensical way of looking at a human being.

Now, however, I realize that the teacher was on to something. As we seek ways to stimulate the economy, we are missing out on a valuable "resource" — human bodies. Today, we have sophisticated extraction procedures, and if we were to purify from our bodies complex components in the form of enzymes, lipids, sugars, nucleic acids and so on, I'm sure we could reap thousands of dollars' worth of molecules. We might even discover new compounds that will fight cancer, help us lose weight, improve memory or postpone aging.

We also can easily keep a body on life-support systems and thus increase its shelf life. So we could continue to harvest blood, hair, antibodies, skin and other useful material for months or even years, which would increase the body's value enormously. For that matter, a body kept in this way could be a valuable test system. We could investigate cancer-causing agents or the harmful effects of cosmetics.

Uteruses could be used as receptacles to study the early stages of embryonic development, while males would yield a steady supply of sperm for artificial insemination.

Of course, the body is a gold mine as the source of organs for transplantation, because, to the recipient, an organ is a precious gift. How precious? Is a kidney worth $20,000? A kidney recipient who no longer requires expensive dialysis can resume a place in the work force and pay taxes, so he or she won't be a drain on the economy. Clearly, the government should be involved in salvaging organs because the economic returns would be enormous. Obviously, the organs and tissues from a single human body could net a return of millions of dollars!

This perverse line of thinking could be extended to determining the best age at which to "harvest" a body, all part of a matter of dollars and cents. However, nothing in such an economic equation would encompass the worth of those organ systems as a whole, namely in a living, functioning human being. A human being has a value beyond anything economic because of factors that simply lie outside monetary considerations.

And is it any different when we try to cost out the "value" of wilderness? The Windy Bay watershed on Lyell Island, for example, is pristine, a remnant of what there once was all along the west coast of British Columbia. It is one of the last of its kind, and when it is gone it will be gone forever.

Can we calculate its value other than through its potential for logging, tourism and fishing? Environmentalists and the Haida people know that there is a worth beyond measure, but as long as they must continue to counter the exploiters with "realistic" economic arguments, we lose sight of the central issue. And we will continue to harvest and destroy because of nature's "resource value." We have to raise the level of discussion of wilderness above jobs and economic growth.

* * *

As negotiations between federal and provincial governments went on, public pressure escalated, but I had to leave Canada on a five-week trip to film in the Soviet Union as the discussions over Windy Bay continued. While I was in Irkuskt, Siberia, I received a desperate call from my wife informing me that the talks had fallen through. The B.C. government had cut off negotiations and decided to make a *provincial* park, thereby allowing logging and mining in the contended area.

Frantically, I wrote the following column, which was then carried by hand to Moscow by one of the members of our crew and from there it was telexed to *The Globe and Mail* in Toronto:

Pain. Grief. Anger. These are my feelings as I write this on the shores of Lake Baikal in Siberia, where Canada seems impossibly remote. I'm trying to make sense of the phone call I received in Moscow three days ago (June 17) from my wife. She told me that British Columbia Premier Bill Vander Zalm has scuttled negotiations with federal Environment Minister Tom McMillan to create a national park of the South Moresby area in the bottom fifteen per cent of the Queen Charlotte Islands. After a decade and a half of rancorous and divisive debate and confrontation between loggers, Haida, environmentalists and government, once again the value of the area has come down to jobs and logging revenue.

Poll after poll has revealed that Canadians list environmental issues as a top priority, one they are willing to pay for in taxes and jobs. During his term in office, former premier Bill Bennett's own pollsters discovered that an overwhelming majority of British Columbians favoured keeping South Moresby as a park.

Federal Environment Minister Tom McMillan has gone a long way with the generous package presented to British Columbia. His final offer was almost three times as much as he told me he had available last April. He has made a heroic effort to preserve the area.

But the issue ultimately is not over money, it is a question of what our society values. Wilderness everywhere is disappearing in response to the global demand for more economic growth and more consumer products. We now point with alarm to other parts of the world at the vanishing tropical rain forests, species extinction and habitat destruction through development. But how can we be so self-righteous when *our* own actions don't match our words?

There is an unseemly haste to start those chain saws again and to get into the Windy Bay watershed, the jewel in the crown of South Moresby and the thorn in the side of the loggers. Already, Frank Bebbin's men, in one of the most cynical moves, cleared the hills flanking Windy Bay while negotiations about the fate of the area were going on. As soon as the loggers go over the top, they will begin the rapid irreversible destruction of a pristine watershed. It will not be done without triggering a massive response from natives and environmentalists.

The number of jobs being "threatened" by the park is fewer than seventy! The federal government has offered a generous package that would more than compensate those loggers for the loss of their jobs. Yet that handful of loggers — minuscule in comparison to layoffs in most other industries — has been able to marshal enormous muscle and outweigh the wishes of the majority of people in the province and the country.

What can we surmise from this? Several things:

1) Frank Bebbin and the logging industry clearly have a powerful "in" with the provincial government. Since Premier Vander Zalm's decision flies in the face of public support, we have to ask what gives a small logging company such political clout.

2) This issue must be ideological. If the park was approved, that is perceived by the logging industry as a "victory" by the natives and environmentalists, rather than as a wise decision made for all people. The B.C. government has thrived on confrontation politics, creating the fiction that preserving South Moresby represents a death threat to the logging industry and fostering the illusion that environmentalists and natives must be enemies of loggers and millworkers.

3) At the heart of the issue is a conflict in "worldviews." One looks at all of nature as a "resource," so that by leaving a "decadent" forest such as the one around Windy Bay, we "waste" it. The other sees wilderness as an immensely complex and balanced ecosystem whose components will never be fully comprehended but whose very existence enriches our lives spiritually. Whatever one's worldview, it is a fact that an ecosystem like the one in Windy Bay is not like Mr. Vander Zalm's Fantasy Garden (a reference to Vander Zalm's private park, which is run by the premier's wife, Lillian), it is the product of thousands of years of evolution and natural selection and can never be duplicated by human beings.

Why is it that virtually every B.C. proposal for federal assistance includes a request for tens of millions for "reforestation"? The answer is that the logging industry's claims that it is "sustaining the yield" of a "renewable resource" is a cruel hoax. Globally, logging industries are plundering the planet and failing miserably to replenish the forests. That is what makes the remaining virgin stands so attractive — they represent much better profits than the so-called "crops" that the industry has sewn.

Once trees like those in Windy Bay are cut down, there is no issue. But until the saws and axes touch those trees, there is time. The federal government has acted with courage and generosity, and all of us who care about showing some humility and leaving something for our children must now register our support for its inititative — massively and unequivocally. Write, telegraph or phone the prime minister, minister of the Environment and your local MP. Register your disapproval to B.C. Premier Bill Vander Zalm and his minister of the Environment. Give money to groups fighting to save South Moresby and be prepared to put your body on the line.

It's time for people to stand up and be heard. Pierre Berton, Farley Mowat, the Nylons, Bob Bateman, you are friends of South Moresby. Speak up now. And what about the B.C. NDP? Where is Mike Harcourt, and what is the position of his party on logging South Moresby? What about the federal NDP and Ed Broadbent? John Turner represents a B.C. riding — what is his position and that of his party? This could be a pivotal issue in a growing political consciousness of the true value of wilderness.

* * *

The above column was never printed. With stunning speed, a massive public outcry arose after Vander Zalm's announcement. It forced a renewal of negotiations and a brand-new federal park reserve was created. The outcome was overwhelming and an all-too-unusual victory for natives and environmentalists alike, so I was able write a rare celebratory column:

It is a moment for all people to savour and for Canadians especially to be proud: the South Moresby area of the Queen Charlotte Islands will be established as a national park reserve to be preserved in perpetuity. This is a splendid achievement with profound implications.

First, we have to congratulate B.C. Premier William Vander Zalm for making a tough and wise decision in agreeing to the park. He negotiated a generous price from the federal government, for which all British Columbians ought to be grateful. And he made a clever strategic move with his initial announcement last month that the province would establish a park within which logging would continue. The ensuing massive response gave him the unequivocal public support for the park that he needed to back his decision to accept the federal offer.

Prime Minister Brian Mulroney's strong support for South Moresby will endure as one of his best political stands. There is no question that without his approval the park would not have come about. And last, but far from least, Environment Minister Tom McMillan must be recognized for his role in creating the park. When I spoke to him last May, he indicated that the most money he could scrounge from a cost-conscious government was about $40 million. Somehow he found the support within Cabinet to increase the offer almost threefold. His refusal to compromise with British Columbia on the park boundaries was vital. At some point, politicians have to make stands on issues that matter more than getting reelected. Tom McMillan did that.

There are others who should be heroic role models for our young-sters. They are people who gave up careers, steady incomes, even relationships to devote a major chunk of their lives to the preservation of South Moresby. Each one deserves an article, but here I can only list them and hope they will not be forgotten: Tom Henley, John Broad-head, Colleen McCrory, Paul George, Bristol Foster, Vicki Husband, Tara Cullis and Jeff Gibbs.

Artist Bill Reid was a spiritual inspiration for all, and the Haida people, led by Miles Richardson, put their culture and land claims on the line and even risked jail to save the area. Through their dedication, these people have left a priceless legacy to last beyond their lives. And I think Canadians should look in the mirror and congratulate them-selves because it was their support that gave politicians the mandate they needed.

What have we learned from all this? South Moresby, particularly Windy Bay, became a symbol for all special wilderness areas. It became a crucial issue because the outcome would indicate what Canadians value and how we see ourselves in relation to nature. The very act of debating the importance of wilderness, logging, tourism, fishing, native rights and mining was part of an educational process going on around the world.

I've just returned from a six-week filming trip in the Soviet Union. The film crew spent eight days at Lake Baikal, Siberia's magnificent treasure that contains as much fresh water as all our Great Lakes combined. For twenty years, Soviet citizens and scientists have been fighting to protect the purity of Baikal. With the help of *glasnost*, they appear to have won.

Logging around the lake has been stopped, a huge girdling park zone established and the paper mill on the lake will be phased out within three years. Earlier this year, after much public pressure, Soviet leader Mikhail Gorbachev finally cancelled a megaproject that had been planned for years. It was a proposal to divert to the south several huge rivers that flow into the Arctic Ocean. It would have caused climatic changes, to say nothing of a massive ecological disruption. (The scheme was as crazy as the idea of converting James Bay into a lake for export of fresh water.) So even in a society as rigid as that in the Soviet Union, environmental values have become an inescapable public issue.

South Moresby does not represent the last skirmish or area to be fought for. Canada's total parks system is minuscule and must be increased so that we can establish a fragile *network of wilderness*, tiny oases of diversity that will be preserved as a hedge against our ignorance and as treasures for our children. We have to invest far more money in silviculture and reforestation. We have to study and copy other countries such as those of Scandinavia that do sustain their logging yields because their forest resources are far more limited than ours.

In the end, what South Moresby revealed was a profound clash between worldviews. The dominant one sees all nature as a potential resource, of value only for its economic worth. But there is growing support for a different outlook that recognizes that we are biological beings who, in spite of science and technology, remain attached to and dependent on nature. We have to fight to keep nature intact and to try to bring ourselves into a balance with the environment. South Moresby could be a watershed that marks a shift towards this re-emerging world-view.

Politicians Weak in Scientific Savvy

*T*he battle over South Moresby showed the power that politicians have and how important it is for them to be scientifically and environmentally aware. Politicians today are making decisions that will determine the world we leave our children. They must be able to see through many of our ungrounded beliefs, yet too often decisions are made for immediate political gains or in ignorance.

If we were to judge the importance of a subject by the amount of coverage in newspapers or on television, we'd have to conclude that the major issues that concern us as individuals and as a society are economic, political, athletic and glamorous. Newspapers devote entire sections to business, entertainment and sports while the main news sections are dominated by political reports.

As political and economic issues are magnified, the things that really matter are de-emphasized or relegated to the back pages. The important factors that are transforming our lives and the planet result from the use of *science* by private industry, medicine and the military. Consider the impact of technologies like computers, nuclear power, oral contraception, space research and genetic engineering and the enormous problems of energy depletion, global pollution, species extinction, habitat destruction, deforestation and desertification. Surely in that context, daily political machinations or scores on the sports pages become trivial. Science and technology are the major factors in the creation of, and part of the solution to, these problems, yet science reporting is generally limited to major newspapers and then only as an occasional subject.

Every one of us should be conversant with science and technology. I don't mean that we should be able to pick up an obscure journal and read a scientific article with comprehension. Nor do I mean that we should know something about all subjects in the vast range of scientific fields. (Scientists themselves are very parochial in the range of subjects they know.) But every citizen ought to understand how science and technology differ from other forms of human thought and activity, what their tools are, how scientists know something, the nature of proof and disproof, what constitutes facts and what are theories. We have to understand some of the basic concepts of matter, ecosystems, heredity and physiology if we are to make informed decisions of importance now and in the future.

Seen from this perspective, our political system seems ill-designed to handle the major issues facing our leaders. Many problems such as acid rain, toxic chemicals in the Great Lakes, reforestation, energy alternatives, fusion energy and biotechnology require immediate action. But the payoffs may come only decades later. No politician in his/her right mind could make a commitment that might have tangible results long after he/she or his/her political party is out of office.

As a science broadcaster, I have met almost every minister of state for Science and Technology. To a person, each was extremely hard-working and sincere, yet each was under incredible pressure from demands on his or her time. Between trips to their constituencies, meeting supplicants and lobbyists, answering mail, attending commit-tee and caucus meetings and other duties, they have little time to read, to think or even to participate in a thoughtful discussion. Even more distressing is the quality of people we're electing. By that I don't refer to their intelligence, commitment, sincerity or integrity, I mean their ability to assess the background information on important matters they'll have to make decisions on. In the fall of 1987, I examined the professions of thirty-eight members of the then current federal cabinet. Of the thirty-two ministers for whom I could infer a profession prior to politics, the breakdown was as follows: law – 12, business – 10, teaching – 5, farming – 3, engineering – 2. Lawyers and business-people were found, a few years ago, to have the lowest comprehension of simple terms and concepts in science and technology, yet they make up seventy per cent of cabinet!

We all tend to rank the priority of issues according to our special interests and backgrounds. Ask a doctor, native person, scientist, welfare recipient, Christian fundamentalist, or policeman for ideas on important issues and you will get radically different answers. We each see the world through the prisms of our own experiences and training, so it's not surprising that our governments tend to be obsessed with matters economic and jurisdictional — because businesspeople and lawyers dominate our legislatures. People from those two professions skew governments' priorities while their scientific illiteracy renders them incapable of making informed decisions about scientific and technological matters.

This is not inconsequential. A government led by people who do not know enough to assess the scientific and technical advice from their experts cannot make wise decisions about whether to stay with the Candu fission reactor or go for fusion, our future in biotechnology, the hazards and benefits of computers or research in outer space. Nor can they fully assess the ecological impact of new technologies, the possi-ble health hazards of radioisotopes, the limitations to environmental assessments. And so decisions are made for political expediency, thereby ensuring that we as a society will back blindly into the future. That's not how we should chart our way into the uncertain times ahead.

We simply cannot afford to be a society that is scientifically uninformed any longer. Anyone who runs for political office must be expected to be literate in science and technology.

Making Waves

The most frequent challenge I encounter as an environmentalist is, Suppose I agree with your analysis. What can *I* do? We are all at different levels of awareness and sensitivity to the environment, but whatever stage we're at, we must educate ourselves and get involved. There's a lot that can be done, and here's a very personal list of suggestions to start with.

Inform yourself. There are many helpful organizations, magazines and books on a wide range of environmental issues. For starters, I'd recommend two old but classic books, Rachel Carson's *Silent Spring* and E. F. Schumacher's *Small Is Beautiful*.

Convince yourself of the reality of environmental degradation. Talk to old people about what the forests, fish, air, birds, mammals were like when they were young. Our elder citizens are a living record of how much the world has changed.

Re-examine some of your most deeply held beliefs. Is steady, continued growth necessary for our well-being? What is the "quality of life"? What is our relationship as a species with the rest of the natural world? What is the ultimate purpose of our government and society? What is progress? Where do spiritual values fit in our lives? The way we answer these questions will determine how we address and act on environmental issues.

Be a conservationist in your daily behaviour. Find out where there are depots for glass, paper, chemicals and metals and start recycling. Use cloth diapers. Store chemical leftovers from the garage, kitchen cupboard or medicine cabinet for proper disposal instead of pouring them into the sink. Compost your kitchen leftovers. You will be amazed at the reduction in garbage.

Use your power as a consumer. Exert pressure by what you do or do not buy. Praise environmentally responsible companies and criticize ones that aren't. Deep Woods, for example, has an excellent pump spray for insect repellent; why should there be any aerosol sprays when a mechanical pump does the job? Urge stores to phase out use of plastics, ask supermarkets to replace foam containers and packaging, demand that fast food outlets abandon the incredibly wasteful packaging that adds nothing to the quality of the food. Make sure hamburger chains use North American beef.

There is no end to profit-driven waste and pollution. Don't buy leaded gas. Point out the energy waste of open upright refrigerators in food stores. Recycle all glassware and metal containers. Don't buy toothpastes simply because they have colour, stripes or a delivery gizmo that add nothing to deter decay. Don't buy breakfast cereals with little nutritional content just because of a packaging gimmick. Buy organically grown food. This list is endless. Share your thoughts and ideas with others.

Exert your influence as a citizen and voter. Urge all municipal governments to start recycling and set a goal of sixty to seventy per cent. No city or town should be allowed to release raw sewage into rivers, lakes or oceans. Boats should not be allowed to dump sewage into salt- or fresh water. We have to begin major initiatives to recycle human sewage onto agricultural fields. Agricultural land should not be used for development or landfill.

Press for more legislation with teeth to stop industrial pollution, tighten regulations and reduce waste. Industry's crocodile tears over excessive costs of pollution control and threatened shutdowns should no longer be allowed to delay implementation of antipollution measures. We need to impose massive fines and jail sentences for corporate executives as well as individuals who pollute. There should be a special police group like the Los Angeles Strike Force to track down and apprehend polluters. Vehicular exhaust should be rigidly controlled. We need massive crash R & D programs on alternative energy, pollution detection and control and environmental rehabilitation. Governments have to approach environmental matters holistically rather than partitioning them into ministerial departments. Politicians act when they feel the heat of public pressure: letters, phone calls and telegrams do have an impact. Praise politicians with good environmental records.

Take an active part in elections. Attend all-candidates meetings and ensure that those running have thought about questions of nuclear energy versus alternative energy, atmospheric degradation, pollution, pesticides, and the relationship between profit and growth and environmental degradation. It should be as essential for any candidate to have a serious environmental platform as to be able to read or add.

Support environmental groups. There are many effective environmental groups fighting at different levels. Contact the Canadian Environmental Network through the federal Environment ministry for the list of organizations across Canada and then choose according to your priorities. They need money, support and volunteer help. My personal list includes Canadian Arctic Resources Committee (CARC), Acid Rain Coalition, Probe International, Energy Probe, Pollution Probe, Canadian Environmental Defence Fund, Greenpeace and the Sea Shepherd Society.

Think about our children. For the first time in human history, we know that our children will inherit a world that is radically impoverished in biodiversity and afflicted with major problems of degraded air, water and soil. Children should learn that pollution and waste are obscene and represent assaults against them. Surely the opportunity for youngsters to anticipate a rich and full life in balance with the complex community of life on this planet is the reason for society and governments. Profit is not the reason.

This list is far from complete. None of the individual acts will save the world from the impending catastrophe, but involvement changes us, provides us with new insights, leads us into different strategies. It's the *process* that matters. It's the struggle that gives hope.

WONDER LOST: HOW EDUCATORS HAVE FAILED

Introduction

*I*n the long run, the only way we can get off our destructive path is to develop a radically different perspective on our place in nature. Adults who have invested a great deal of effort to reach their current positions are reluctant to make that shift. *Children*, on the other hand, have most at stake since they will inherit the world we leave them; since they have not invested the time and effort in the status quo, they are still receptive to possibilities and options. So we must fight to save as much wilderness as possible everywhere while simultaneously working to ensure that our children are different from us.

I took my children to the zoo recently and it brought back memories of my first visit to a major zoo. My trip to the Detroit Zoo in the late forties is burned into my memory as a high point in my childhood. I was overwhelmed with the astonishing diversity of life, and for years after, I dreamed of travelling to study animals in exotic places where life flourished. Back then, Africa was still a mysterious continent reputed to have areas where white people had never been. The Amazon River was still dominated by headhunters and impenetrable jungle. And there remained the sense of the vast limitlessness of animal and plant species.

What a difference forty years makes. At every zoo exhibit, my seven-year-old daughter would ask me, "Are there many of these left?" And each time, I found myself answering, "No, these are endangered" or "These are disappearing." I had to tell her the magnificent California condors have been reduced to a handful and will be gone before she's a teenager. I reported that there are more rhinoceroses in the Cincinnati Zoo than live on the entire Serenghetti Plains, that cheetahs and Siberian tigers may soon join a growing list of animals extinct in the wild. Instead of being an opportunity to revel in nature's richness and abundance, a visit to the zoo has become a time of sorrow!

"Why is everything rare?" Severn kept asking. Why, indeed? A child's innocent question penetrates to the dilemma of our times.

When I lived in Leamington, in the forties, every year there was a massive hatch of mayflies. Those insects would emerge from Lake Erie as adults in numbers that defied description. Cars would skid on their bodies on the road, and they would completely envelop houses and darken streetlights. On the beaches, mayfly carcasses would form piles over a metre thick. Those insects fed countless fish, mammals, birds, other insects and microorganisms. It was a spectacular example of nature's abundance at the base of the food chain.

Today, farm runoff, industrial pollution and sewage have wiped out that immense hatch. People shrug and call it "the price of progress." Perhaps they're glad they don't have the nuisance of putting up with all those "messy" flies. We may have got rid of those pesky flies, but at what cost to the ecosystem? We are changing the world around us beyond recognition with consequences that are totally unpredictable. The signs are unmistakable — we are creating a world that is not only unlivable for plants and animals but also for us.

When I grew up in the forties, we used many times less of everything — energy, space, air, water, resources — yet we lived rich, full and meaningful lives. This is not romantic nostalgia that comes with age. Within the lifetimes of our elders, the environment around us has been radically transformed.

Disappearing rhinos, mayflies and old-growth forests are a barometer of what we are doing to the planet, and all the indications are of a massive danger that dwarfs questions about Free Trade, for instance, to insignificance. The real challenge we face is to redefine "progress."

I left the zoo the other day with all these heavy thoughts. But it didn't work that way for my daughters. I got home from work three days later to find them with their friends and their parents parading up the street carrying signs that said "Save the Wildlife! Write your MP!" Children can see the obvious — that their world is being steadily denuded. Why can't grown-ups?

Teaching the Wrong Lessons

*C*hildren learn by example. They watch parents and quickly pick up attitudes from our actions. In spite of the vast expanse of wilderness in this country, most Canadian children grow up in urban settings. In other words, they live in a world conceived, shaped and dominated by people. Even the farms located around cities and towns are carefully groomed and landscaped for human convenience. There's nothing wrong with that, of course, but in such an environment, it's very easy to lose any sense of connection with nature.

In city apartments and dwellings, the presence of cockroaches, fleas, ants, mosquitoes or houseflies is guaranteed to result in the spraying of insecticides. Mice and rats are poisoned or trapped, while the gardener wages a never-ending struggle with ragweed, dandelions, slugs and root rot. We have a modern arsenal of chemical weapons to fight off invaders, and we use them lavishly.

We worry when kids roll in the mud or wade through a puddle because they'll get "dirty." Children learn attitudes and values very quickly, and the lesson in cities is very clear — nature is an enemy: it's dirty, dangerous or disgusting. So youngsters begin to wall themselves from nature and to try to control it. I am astonished at the number of adults who loathe or are terrified by snakes, spiders, butterflies, worms, birds — the list seems endless.

Yet for ninety-nine per cent of our species' existence on the planet, we were respectful of and dependent on nature. When plants and animals were plentiful, we flourished. When famine and drought struck, our numbers fell accordingly. We remain every bit as dependent today; we need plants to fix photons of energy into sugar molecules and to cleanse the air and replenish the oxygen. It is folly to forget our dependence on an intact ecosystem, but we do whenever we teach our offspring to fear or detest the natural world. The message urban kids get runs completely counter to what they are born with, a natural interest in other life forms. Just watch a child in a first encounter with a flower or an ant — there is instant interest and fascination. We condition them out of it.

I see it when my ten-year-old daughter brings home new friends and they recoil in fear or disgust as she tries to show them her favourite pets — three beautiful salamanders. And when my six-year-old comes wandering in with her treasures — millipedes, spiders, slugs and sowbugs that she catches under rocks lining the front lawn — children and adults alike usually respond by saying "Yuck."

I can't overemphasize the tragedy of that attitude. For inherent in this view is the assumption that human beings are special and different and that we exist outside nature. Yet it is this belief that is creating many of our environmental problems today.

As long as we have cities and technology, does it matter whether we sense our place in nature? Yes, for many reasons, not the least of which is that virtually all scientists were fascinated with nature as children and retained that curiosity throughout their lives. But a far more important reason is that if we retain a spiritual sense of connection with all other life forms, it can't help but profoundly affect the way we act. The yodel of a loon at sunset, the vast flocks of migrating waterfowl in the fall, the indomitable salmon returning thousands of kilometres — these phenomena of nature have inspired us to create music, poetry and art. And when we struggle to retain a handful of California condors or whooping cranes, it's clearly not from a fear of ecological collapse, it's because there is something obscene and frightening about the disappearance of another species at our hands.

If children grow up understanding that we are animals, they will look at other species with a sense of fellowship and community. If they understand their ecological place, the biosphere, then when children see the great old-growth forests of British Columbia or the Amazon being clearcut, they will feel physical pain because they will understand that those trees are an extension of themselves.

When children know their place in the ecosystem and see factories spewing poison into the air, water and soil, they will feel ill because someone has violated their home. This is not mystical mumbo-jumbo. We have poisoned the life-support systems that sustain all organisms because we have lost a sense of ecological place. Those of us who are parents have to realize the unspoken, negative lessons we are conveying to our children. Otherwise, they will continue to desecrate this planet as we have.

It's not easy to avoid giving hidden lessons. I have struggled to cover my dismay and queasiness when Severn and Sarika appear holding a large spider or when we've emerged from a ditch covered with leeches or when they have been stung accidentally by yellowjackets feeding on our leftovers. But that's nature. I believe efforts to teach children to love and respect other life forms are priceless.

Where Are All the Women?

*I*t has been my experience that women are disproportionately represented in the environmental movement, and when I give talks, it is women who get my message right away. I'm convinced it's because they have been excluded from much of the power structure and game playing in society and since they don't have the same stake in the status quo as men they can see through our social myths with greater clarity. And with a much more direct involvement in the nurturing of children, women have a longer perspective on the world we leave to them. But the exclusion of females from science and technology means that they do not bring an important perspective to the most powerful disciplines in society.

In 1969, I taught genetics at the University of California at Berkeley. One of the brightest graduate students in the department was a woman studying population genetics. She had received a bachelor's degree in physics, but when she had applied for graduate school in the subject, she was told that women were not encouraged in the field! Eventually, physics' loss was genetics' gain, and she is now a professor in an Arizona university. I hope that her experience would be shocking and unacceptable today.

These days in the United States, legislation has put pressure on universities to recruit female faculty members, and there has been a corresponding increase in opportunities for women in graduate schools. But there is still a long way to go, especially in Canada. I became a faculty member in the zoology department at the University of British Columbia in 1963. At that time, at the level of assistant professor and above, there were twenty faculty members, none a woman. Today, we have expanded to thirty-eight but only one is a woman. The numbers in other departments aren't much different. Here are the numbers of faculty members in each science department at UBC, followed by the number who are women: Botany – 26, 3; Microbiology – 13, 2; Physics – 55, 1; Chemistry – 49, 0; Mathematics – 60, 1; Geology – 21, 1.

I'm sure these sex ratios are not atypical of university departments across Canada. There is simply a dearth of women in the sciences at the faculty level, with the consequent paucity of role models to attract and inspire women students. Of course you may ask, does it matter that there are so few women in science? It sure does.

Science is a highly competitive, macho activity that exacts a heavy price from many of its predominantly male participants. It is vital for men to get a very different perspective and attitude that could come from large numbers of women in science. I'm sure different kinds of questions would be asked and interpretations of data would differ if there were more women in science. Science and scientists would be better off for their presence. But how do we recruit more females?

I sat on a committee that reviews requests for grants to increase public awareness of science. Many of the applications came from university women's groups that proposed to make films about women in science, to arrange special days for female students to meet women scientists and learn about opportunities for them, to build displays and to have material printed up — all exhorting girls to go into science. This is all well and good, but my own feeling is that such information comes far too late to make a significant change in the numbers of girls choosing a career in science or even to change their attitudes to it.

My opinion is based on personal observation. I have two young daughters, and not surprisingly they reflect a lot of the idiosyncrasies of their parents. When I was four, I caught a salamander while fishing for trout, and for years it was one of my most prized pets. It's no surprise that Severn has a collection of salamanders and newts that she and her grandfather caught, and they have travelled with us back and forth between Vancouver and Toronto at least six times.

Severn feels the same way about her salamanders as I did about mine. What is shocking to me is that many of the girls she has brought over wince and shrink away when Severn dips in and pulls out her amphibians. They say things like "Take them away!" "They're so yucky!" "I hate them!" or "They might hurt me!"

That is not an instinctive, genetically programmed reaction — it is learned! Even little boys, especially city residents, as I've described, are alienated from nature, often treating creatures like frogs and salamanders as creepy crawlies that can be tormented and abused. But it is the girls who exhibit the most stereotypical reactions.

Sarika, my youngest, loves to turn over rocks and pick up sowbugs (those little flat creatures with lots of legs — the ones that roll up into a ball when frightened). She'll also grab millipedes, worms, beetles, slugs and centipedes. (Ever since I was a child and got pinched by a large centipede, I've never liked them, but I'm delighted that I have hidden that from Sarika.) But again, when toddlers come by, they are already conditioned and they react to her collections with fright or revulsion. Fortunately, curiosity quickly overcomes that initial reticence and I usually find them converted (often to the horror of their parents) and soon racing Sarika to collect their own critters.

In grade 1, Severn had a wonderful teacher who, by her enthusiasm, showed a profound love of nature. On her own, the teacher collected milkweed plants and a couple of monarch butterfly caterpillars. She was as delighted and thrilled as the students when the caterpillars pupated to form lovely green chrysalids and then emerged as butterflies a week later.

She was delivering a number of messages to the class — through her enthusiasm, she was showing them the real foundations of interest and inquiry that underlie science; as a woman, she was providing an important role model for the girls; and she didn't make a big point of mystifying science by saying, "This is SCIENCE."

Sadly, she is rare in the school system, and by the time girls finish high school, science will be an activity from which most will have been excluded — not as overtly as the physics student at Berkeley, but every bit as effectively.

I believe it's at the preschool and elementary levels that attention is most desperately needed to stop the pernicious social conditioning that squelches the natural curiosity and interest of all children, but especially girls.

Losing Interest in Science

*T*he most important issues that the next generation will have to contend with will result from the application of science and technology. An interest in these fields needs to be encouraged. Unfortunately, for many youngsters today, the way science is taught in school turns them off, and too many have stopped taking any science courses midway through high school. It doesn't have to be that way.

When my daughter Severn was five, she asked me to explain how a plant's root had the strength to grow through the ground. Looking out the window, I remarked that the long bump in the road was caused by the tree's roots tunnelling under the pavement and pushing it up. Well, this amazed her, and in that instant I re-experienced a moment of discovery that each of my children has shared with me.

Every one of my children has grown up with the weekend nature hike, first in the large woods adjacent to our home, and then along the seashore. All my youngsters and their countless friends loved these outings — I have yet to meet a child who doesn't.

They very quickly learn where to hunt for salamanders, the edible mushrooms, robins' eggs or wood-boring insects. They know from the frogs' songs when to go and look for eggs so they can follow metamorphosis. They never have to be told that what they're learning is entomology, developmental biology or mycology — they just love it for what they get out of it.

And they ask questions that, as a biologist, I find fascinating, questions like: "Daddy, why are your eyes so small?" "Why is Sarika's skin darker than mine?" "Why does Lisa's daddy talk funny?" (he has a cleft lip and palate) or "Can a mommy dog ever given birth to a baby kitten?" These are the kinds of questions that have intrigued many biologists for their entire professional careers and that got many of us into science in the first place.

What happens to that innocent joy in learning about the world going around? As educators design new curricula, I would plead with them not to perpetuate any longer one of the great myths and turnoffs — *the scientific experiment.*

When I was a high school student, we went into a lab and were told what the experiment would be, received a set of instructions and then were expected to use the equipment to obtain data. Because the emphasis was on the mechanics of doing experiments, we frequently lost sight of the reason for doing them. Without an appreciation of the body of knowledge, insights and theories that make an experiment definitive, a student can go through a lab exercise like a cook following a recipe.

In high school, the part of a lab exercise most prized by teachers seemed to be the "write-up." We were drilled in the proper steps: define the purpose, describe the materials and experimental methodology, document the results, discuss the implications and finally draw conclusions. Not only did we have to conform to this protocol, but our reports were often graded on whether we obtained the "right answer."

Having been a scientist now for more than twenty-five years, I can tell you that this is *not* how science is done, and we lose a great deal by teaching it this way. To begin with, there is no such thing as a right or wrong answer — if we knew the right answer beforehand, we wouldn't bother doing research. But even when we repeat a test that has been well documented, the data we get are not "wrong" if they fail to conform to expectations. We may have duplicated the experiment poorly, but the data are all one has.

Those school exercises gave the impression that scientists attempt to solve a problem by proceeding in a linear fashion from A to B to C to a solution. In my opinion, it works much more like my experience in genetics.

In the late 1960s, my lab had demonstrated that it is possible to recover mutations in fruit flies that are temperature-sensitive. Thus, at one temperature such mutants are indistinguishable from normal flies, whereas just a few degrees warmer or cooler, they die or exhibit a wide array of defects. This sensitivity is a useful property; by controlling temperature, we can determine when and where the defective gene is acting during the fly's development from an egg to adult.

In bacteria and viruses, such mutations represent a single molecular change in the gene product that makes it less stable at certain temperatures. How could we find out whether this was the basis for temperature-sensitivity in fruit flies?

We decided to look for a protein defect and focussed on muscle, which flies have lots of. What effect would a temperature-sensitive defect in a muscle protein have? We speculated that it might be flexible and permit movement at one temperature, but "seize up" and cause paralysis at another. So we chose to look for flies with the trait of paralysis. Then the questions came flooding in. Do we look for non-movement in larvae or adults? How do we pick out the paralytics from mobile flies? How would we induce mutations? How would we impose the temperature regime? How many flies would we have to screen? Each question had to be answered by trying small experiments to guide us in the proper direction. We had to build special equipment, design genetic combinations and expand our cultures for a huge test run. All this took months and we were driven to do it by the amused skepticism of our colleagues. Once we got going, we recovered many flies that were dead or sterile or failed to transmit the paralytic condition to their offspring.

After screening more than a quarter million flies carrying chromosomes exposed to powerful chemical mutagens, we recovered a temperature-sensitive paralytic mutation. At 22° Celsius, the flies could fly, walk, mate — they seemed normal — but when shifted to 29° C, they instantly fell down, completely paralyzed. When they were placed in a container kept at 22° C, they were flying again before they hit the bottom! It was a spectacular mutant, and when we found it we screamed and danced and celebrated.

But what was its molecular basis? We soon discovered that our mutation (and many others that were subsequently recovered) was a defect in *nerves* rather than muscle.

And how did we write up our results? We riffled through all our records, selected the ones that said what we wanted and then wrote the experiment up in the proper way: purpose, methods and materials, results and so forth. The paper was written in a way that suggested that we began with a question and proceeded to find the answer. That's because the report was a way of "making sense" of our discovery, of putting our results into a context and communicating it so colleagues could understand and repeat (if necessary) what we did. But it conveyed nothing of the excitement, hard work, frustration, disappointment and exhilaration of the search — or the original reason we started the search for paralytic mutants!

By emphasizing a proper way to do an experiment and to write it up, we create a myth about how research is done. And we lose all the passion that makes the scientific enterprise so worthwhile. I hope the new science curricula don't make this mistake.

Right now, science is being taught on a totally unrealistic model and unfortunately, for the majority of our students, it's a turnoff very early. Indeed, the word *science* is pretty much of a pejorative by the time they reach high school — it's a subject for the "math brains." And certainly for most teenagers, science is an activity so esoteric that it really isn't relevant to their daily lives. What a sad state, when sex, drugs and jobs are important to them, and science has a lot to contribute to an understanding of these issues.

The vast majority of teachers who teach the pitiful amount of science in primary schools is very poorly grounded in science, having had perhaps a few hours of lectures in the education faculty a decade or more previously. It's certainly not the teachers' fault, but these days when we hear so much about the information explosion and the need to get in on the action in high technology it's tragic that so many children are uninterested in science by the time they reach more highly qualified teachers in high school.

That natural capacity to be excited when discovering things in the world around is so precious and so easily extinguished that I think political posturing about getting Canada into world-class science is a waste of time unless we devise ways to keep our most talented youngsters interested.

The students I teach in the university are all enrolled for science degrees. They are committed, with specific jobs in mind — medicine, forestry, agriculture — and sadly, most of them have already lost that sense of wonder and joy in learning. They have managed to survive the science curricula but have been pounded into submissive memorizers and grade grubbers. When I returned to the university in 1985, I stopped teaching in classes.

I had always prided myself as a teacher. Teaching to me was a very intimate thing: it meant sharing a part of my life, my thoughts with the class. In the early eighties I felt brutalized by the students' preoccupation with what would be covered on the exam, what percentage of their final grade this material would make, and so on. I feel too old to put up with that anymore.

But I have two youngsters who, like their old man, are crazy about

their salamanders, their beetle collection and their pressed flowers. I
don't want to see them lose that. Something must be done about science
in the schools.

God in the Machine

*S*peaking of computers — and who isn't thinking or
talking about them today — isn't it amazing how quickly our attitudes
towards them have changed? When I first started to report on computers on radio and television about twenty years ago, all the emphasis
was on explaining how computers work. These days, the innards of the
microelectronic revolution are of little concern. All our attention is
focussed on *what* they can do. That's fair enough. In the beginning,
you had to have a sense of what the limits and potential of the
technology were because there was a good chance you might have to
fix it yourself. But now there are all kinds of specialists for you to get
help from, and it's no more important to know how the computer works
than it is for the average driver to understand how a car's automatic
transmission operates.

It's very clear that computers are here to stay and we're already
oblivious to them in our social landscape. At airports we take advance
seat selection for granted. We can get boarding passes for connecting
flights a continent away, something impossible before computers. We
accept that when "the computer is down," we have to wait because
human workers can no longer do things by hand. My wife, who teaches
writing at Harvard, reports that the most frequent excuse for missing a
deadline for a paper is problems with the computer. In my day, it was
sickness or personal problems.

Only a couple of years ago, parents were panicked by predictions
that their children would probably require computers in their jobs in
the future. Parental hangups and fears about computers led many to
send their children to summer computer schools, to buy computers for
their kids and to lean on schools to get more computers. That cry for

computer literacy has been, in my view, one of the biggest cons ever foisted on the school system.

First of all, schools should not get into the position of trying to anticipate the future needs of the workplace. If history is any guide, our ability to predict future requirements becomes very fuzzy beyond a couple of years. But it is not the obligation of schools to prepare youngsters for the job market, at least not beyond the universal skills of thinking, reading, calculation and so on.

There are those who see in the computer a means for teachers to teach children how to think through a series of problem-solving challenges. Nonsense. That is a human activity demanding human guidance and understanding. Nothing will ever replace the flexibility and originality of the human mind. No machine will ever duplicate a good teacher. And do we want to involve our children in what is basically a solitary activity at a time when they should be acquiring social and language skills?

Many of the people who now look to the computer as the great gift to education are the very same ones who were trumpeting the promise of video technology a decade ago. They are technocrats who jump on the superficial benefits without consideration of the broader ramifications or costs.

In our rush to ensure that our children are computer literate, we have forgotten to ask what we are trying to accomplish in the schools. After all, the computer is simply a machine. Is it any different from an automobile or television set? We didn't change the entire curriculum of schools to accommodate those technologies when they were new. Nor would we think of training our youngsters to drive a car using a Model A Ford, yet in terms of what today's youngsters will use when they grow up, present-day computers will seem pretty crude indeed. Back in the 1960s, I struggled to learn Fortran, the computer language of the time, and I can tell you that the ease with which we now use computers staggers me. Computers will continue to get easier and easier to use.

So what's the panic? Children will grow up accepting computers as a part of the landscape of their culture and will use them as the need arises. The Ontario Ministry of Education has embarked on an ambitious program to try to make all teachers in that province computer literate. This has all the earmarks of the ill-fated program to make all civil servants bilingual. What's the point? The young teachers coming along will all take computers for granted. Teachers who find a real pedagogical use for computers will overcome their reticence. If they see no use for a computer, why force it on them?

The computer provides access to vast amounts of information with remarkable speed. But one of the biggest problems educators face today is that youngsters already have an overload of information, especially from television. Is the computer's faster access to even more information going to help the teacher? No. The challenge is to decide what information is worth anything and how to make use of that information. That is taught best by teachers. I think we are misguided in our panic about computers. And diverting scarce funds for primary school teaching into such technology only exacerbates the problems we already have dealing with technology.

Mad Doctors, Mad Machines

*T*oday the omnipresence of the computer is undeniable and its proponents are legion. But the provocative commentator on society Theodore Roszak has offered profound reasons for objecting to the infatuation with computers in education.

Roszak has been writing for years about the place of technology in society. As a scientist, I was electrified by one of his passages in a short essay:

Dr. Faustus, Dr. Frankenstein, Dr. Moreau, Dr. Jekyll, Dr. Cyclops, Dr. Caligari, Dr. Strangelove. The scientist who does not face up to the warning in this persistent folklore of mad doctors is himself the worst enemy of science. In these images of our popular culture resides a legitimate public fear of the scientist's stripped-down, depersonalized conception of knowledge, a fear that our scientists, well-intentioned and decent men and women all, will go on being titans who create monsters . . . . What is a monster? The child of knowledge without gnosis, of power without spiritual intelligence.

Roszak's books, which include *Unfinished Animal*, *Where the Wasteland Ends* and *The Making of a Counterculture*, are disturbing and provocative. His latest, *The Cult of Information*, is a penetrating look at computers. One of his warnings about the computer revolution is the devaluation of *ideas* by a new fad — *information*. He writes:

> For the information theorist, it does not matter whether we are transmitting a fact, a judgment, a shallow cliché, a deep teaching, a sublime truth, or a nasty obscenity. All are information. The word has gained a vast generality of meaning, but at a price: things communicated come to be levelled, and so too the value. The main concern of those who use information theory is with apparatus, not content. We live in a time when the technology of human communications has advanced at blinding speed, but what people have to say to one another by way of that technology shows no comparable development . . . .
>
> Depth, originality, excellence, which have always been factors in the evaluation of knowledge, have somewhere been lost in the fast, futurological shuffle . . . this is a liability that dogs every effort to inflate the cultural value of information.

He suggests that there is a powerful message conveyed with the introduction of computers in education:

> Introducing students to the computer at an early age, creating the impression that their little exercises in programming and game playing are somehow giving them control over a powerful technology can be a treacherous deception. It is not teaching them to think in some scientifically sound way; it is persuading them to acquiesce.
>
> It is accustoming them to the presence of computers in every walk of life, and thus they become dependent. The best approach to computer literacy might be to stress the limitations and abuses of the machine, showing students how little they need it to develop their autonomous powers of thought.

Sherry Turkle, a psychologist and sociologist at the Massachusetts Institute of Technology, observes that, in time past, children learned their human nature in large measure by comparing themselves to animals. Now, computers with their interactivity, their psychology, with whatever fragments of intelligence they have, bid to take this place.

It would indeed be a loss if children failed to see in the nesting birds and the hunting cat an intelligence as well as a dignity that belongs to the line of evolutionary advance from which their own mind emerges.

How much ecological sense does it make to rush to close off what remains of that experience for children by thrusting still another mechanical device upon them? Science and technology at their highest creative level are no less connected with ideas, with imagination, with vision.

It would surely be a sad mistake to intrude some small number of pedestrian computer skills upon the education of the young in ways that blocked out the inventive powers that created this astonishing technology in the first place. And what do we gain from any point of view by persuading children that their minds are inferior to a machine that dumbly mimics a mere fraction of their native talents?

To Roszak, it is a terrible mistake to look to the computer as a panacea in education:

> Free human dialogue, wandering wherever the agility of the mind allows, lies at the heart of education. If teachers do not have the time, the incentive, or the wit to provide that, if students are too demoralized, bored, or distracted to muster the attention their teachers need of them, then *that* is the educational problem that has to be solved.

And the ultimate effect of computers in the early years of schooling might be: "We may soon be graduating students who believe . . . that thinking is indeed a matter of information processing, and therefore without a computer no thinking can be done at all. Defaulting to the computer is not a solution; it is surrender."

The Cult of Information asks uncomfortable questions about our infatuation with the spectacular advances in computer technology.

Roszak believes: "There can be the real danger that we fall prey to a technological idolatry, allowing an invention of our own hands to become the image that dominates our understanding of ourselves and nature."

To support the reality of this hazard, he quotes Princeton physicist Robert Jastrow, who has said, "We can expect that a new species will arise out of man, surpassing his achievements as he has surpassed those of his predecessor, *Homo erectus*. The new kind of intelligent life is more likely to be made of silicon."

This is Roszak's retort to Jastrow:

Have we created something like a mind, one that is better suited to the alienating conditions of modern society, better able to cope with the pressure, the anxiety, the moral tension? If that were so, it might be taken as a damning judgment upon the inhumanity of the social order we have created.

But some computer scientists clearly regard it, instead, as an indictment of human nature itself; we possess a mind which is not fit to survive . . . . It can lead to the self-serving argument that more power should be entrusted to the machines that computer scientists have invented and control.

In order to follow Roszak's warnings about the computer, we have to understand what he regards as the difference between a human brain and a computer:

We ordinarily take in the flow of events as life presents it — unplanned, unstructured, fragmentary, dissonant. The turbulent stream passes into memory where it settles out into things profoundly remembered, half remembered, mixed, mingled, compounded. From this compost of remembered events, we somehow cultivate our private garden of certainties and convictions, our rough rules-of-thumb, our likes and dislikes, our tastes and intuitions and articles of faith.

Memory is the key factor here; it is the register of experience where the flux of daily life is shaped into the signposts and standards of conduct . . . . Computer memory is no more like human memory than the teeth of a saw are like human teeth; these are loose metaphors that embrace more differences than similarities . . . . The cult of information obscures this distinction, to the point of suggesting that computer memory is superior because it remembers so much more. This is precisely to misinterpret what experience is and how it generates ideas.

Human memory is the invisible psychic adhesive that holds our identity together from moment to moment. It flows not only through the mind, but through the emotions, the senses, the body.

Our memory is rigorously selective, always ready to focus on what matters to us. It edits and compacts experience, represses and forgets. In the human mind, an original idea has a living meaning: it connects with experience and produces conviction. What a computer produces is "originality" at about the level of a muscular spasm.

And the outpouring of data has great political significance.

We suffer from a glut of unrefined, undigested information flowing from every medium around us . . . . There can be *too much* information. Information is transformed into a political issue when it is illuminated by an idea — about justice, freedom, equality, security, duty, loyalty, public virtue.

We inherit ideas like these out of our rich tradition of political philosophy: from Plato and Aristotle, Machiavelli and Hobbes, Jefferson and Marx. Very little of what these minds offer is connected with information. If it were, the data would long since have been antiquated; but the ideas live on. Facts, if they are to be of any value, need to be used in the service of images and ideas like these. We would be better off if the mainstream public were actively in touch with a few good journals of opinion (left, right, and centre) than if we had a personal computer in every home.

The computer lends itself all too conveniently to the subversion of democratic values — the ability to concentrate and control information. All that the data banks and their attendants are after is information at the most primitive level: simple, atomized facts. For the snoops, the meddlers, data glut is a feast. It gives them exactly what their services require. They exist to reduce people to statistical skeletons for rapid assessment.

The most obvious group that manipulates this information, in Roszak's opinion, consists of the pollsters. Polling is a cause of a degradation of politics:

> The candidates *think* the pollsters make a difference and plan their campaigns in response to computerized information. The result is a dismal style of politics that grows more obsessively focused on imagery, sloganeering, rhetorical legerdemain — in fact, on the huckstering skills of the marketplace. Pollsters only ingrain the corruption deeper by claiming they can manoeuvre their candidates with pinpoint accuracy.
>
> The transient trivialities that register in computerized polling . . . have nothing to do with thought or conviction; . . . the electorate is placed in the absurd position of becoming spectators to its own predicted political behavior.

Roszak forces us to ask some fundamental questions about what is unique about the human brain and whether we are making a mistake in keeping the computer in such high regard. But it is the relationship of computers with the military that Roszak finds most chilling:

> If the measure of political power is the magnitude of the decisions one makes, then we may be only inches away from entrusting the computer with our society's supreme governing authority.
>
> Both in the United States and the Soviet Union, more and more control over the world thermonuclear arsenal is being lodged in computerized systems. This is nothing less than the power over our collective life and death.

Roszak points out that because of the speed and accuracy of missiles, in the event of an attack, "the response time on the part of the United States would be so short that the missiles would have to be fired at the earliest possible moment."

He notes that the computer control of messages to and from military command centres hasn't always worked:

> In 1977, during a systematic test, more than 60 per cent of its messages failed to get transmitted. During one 18-month period in

the late 1970s, the North American Air Defence command (NORAD) reported 150 "serious" false alarms, four of which resulted in B-52 bomber crews starting their engines.

As the response time allowed by the weapons grows shorter, the system will have to become more hair-triggered and autonomous. The Pentagon is presently seeking to develop an artificial intelligence expert system that might be placed in full control of the weapons . . . .

The weapons will lie wholly in the province of machines whose artificial intelligence will be an unfeeling, one-dimensional caricature of soldierly will. The final most decisive act of war conceivable will come down to the logic of numbers processed coldly and swiftly through the inscrutable silicon cells of a machine.

War, insurrection and public unrest around the world are currently "managed" by experts:

The crisis managers . . . purport simply to be providing objective information on demand. But this information has been edited, shaped, weighted, and ordered; it reflects priorities perhaps of a highly ideological kind.

The more simplified and graphic the computerized representations become, the more densely interpretive they will have to be. Thus, the graphics that were presented to President Reagan by his crisis managers during the Grenada invasion must have been shot through with assumptions about Caribbean politics, worldwide Communist intentions, America's role in the Western Hemisphere.

The programmers are likely to present a facade of professional objectivity and uniformity. Nevertheless, they are fashioning an intellectual artifact that reflects the tastes, choices, judgments of its makers.

Roszak quotes a horrifying observation about military supercomputers from computer expert Joseph Weizenbaum of MIT:

These often gigantic systems are put together by teams of programmers, often working over a time span of many years. But by the time the systems come to use, most of the original programmers have left or turned their attention to other pursuits. It is precisely when gigantic systems begin to be used that their inner workings can no longer be understood by any single person or by a small team of individuals.

Roszak adds:

Machines that are run by incomprehensible programs are approaching a kind of technological madness. They are on their way to becoming disintegrated intelligences, split and fragmented, a psychotic stew of assumptions and standards that can no longer be given a rational order.

Yet, if the machine is part of a crisis-management project, it will continue to process data, packaging it neatly into charts, graphs, simulations, giving the surface appearance of rationality.

Computers are gifted with such a facade of impersonal precision; they have no way to look or sound or act crazy. Of course, the true madness would be that of the decisionmakers and the society at large that elected to become dependent on mechanical systems that bring these liabilities with them. Just as the ultimate proof of the mechanistic model in medicine is the invention of a mechanical heart or kidney that will sustain life, so in cognitive science the effort is to invent a machine that will convincingly imitate the highest functions of the mind — its power to reason, to judge, to decide.

The long-range dangers of all this lie in the equation of information with ideas, because:

. . . an information economy spends more and more of its resources accumulating and processing facts. We bury ever deeper the substructures of ideas on which information stands, placing them further from critical reflection.

We begin to pay more attention to "economic indicators" than to assumptions about work, wealth, and well-being which underlie economic policy.

The hard focus on information that the computer encourages must in time have the effect of crowding out new ideas, which are the intellectual source that generates facts.

There are a lot of people who will dismiss this as so much rubbish. But if we can suspend that unthinking rejection and consider what Roszak is saying, he raises questions about technology that go far beyond the computer.

Why Johnny and Mary Can't Experiment

*P*re-eminence of the United States in industry and technology has been severely eroded by upstarts from the Orient. And just as they did after Sputnik, Americans are trying to figure out why. In part, the answer is simple. Only thirty years ago, American goods made up sixty per cent of world trade and today they've dropped to about twenty-five per cent. The catastrophic decline for the U.S., which has only six per cent of the world population, is merely a more equitable global adjustment. But in the search for a culprit, Americans are focussing once again on education and pointing to a failure to teach science adequately.

In 1983, a U.S. National Science Board committee reported:

Alarming numbers of young Americans are ill-equipped to work in, contribute to, profit from and enjoy our increasingly technological society. Far too many emerge from the nation's elementary and secondary schools with an inadequate grounding in mathematics, science and technology. As a result, they lack sufficient knowledge to acquire the training, skills and understanding that are needed today and will be even more critically needed in the 21st century.

Since that study, more than one hundred reports have been made, each calling for reform in science education including updating standards, strengthening teacher preparation, more use of assessments, more research into teaching. This September, the National Assessment of Educational Progress (NAEP) issued a devastating summary of a study of nine-, thirteen- and seventeen-year-olds carried out from 1969 to 1986. Titled *The Science Report Card*, it concludes that student performance in all three age groups *declined* after 1969 and has never again reached the 1969 level. The report finds that more than half of seventeen-year-olds are poorly equipped to handle science courses, participate in a technological issue of social relevance or perform in the workplace.

The NAEP summarized another study by the International Association for the Evaluation of Educational Achievement that compared student performance in science in fifteen countries. It found "At grade 5, the U.S. ranked in the middle in science achievement relative to 14 other participating countries. At grade 9, U.S. students ranked next to last. In the upper grades of secondary school, 'advanced science students' in the U.S. ranked last in Biology and performed behind students from most countries in Chemistry and Physics." Although it is questionable whether student populations in different countries are completely comparable, American educators are understandably alarmed.

The NAEP report documents the enormous gap in science performance by black and Hispanic students relative to whites: "average proficiency of 13- and 17-year-old black and Hispanic students remains at least four years behind that of their white peers." While nine-year-old boys and girls perform equally in science, by thirteen and seventeen, a decrease in female proficiency appears. In looking for causes, the report points to the lack of role models for minority members and girls, socioeconomic disadvantages and social pressure.

Biology is the preferred science for high school students (ninety per cent of students took at least one course), forty-five per cent had chemistry and twenty per cent physics. Yet only six per cent of high school students in the U.S. take an advanced biology course, compared to forty-five per cent in Finland and twenty-eight per cent in English-speaking Canada. NAEP estimates that only seven per cent of American high school students have the knowledge and skills to handle college-level science courses.

Student performance in science is also a reflection of the low priority of science in the curriculum. Two-thirds of grade 3 teachers spend two hours or less a week on science, while less than half of grade 11 teachers spend three hours or more. Only a third of grade 7 and half of grade 11 students are asked to hypothesize or interpret data in science class on a regular basis. Over half of third graders and eighty per cent of eleventh graders never go on science field trips. Many don't even have access to science labs.

As in Canada, education in the U.S. is controlled by the state and there are enormous differences in the quality of education from state to state and county to county. There are over sixteen thousand different school boards, each wielding much power. And the problems will increase when, by the year 2000, one in every three Americans will be nonwhite. There will be more students who are poor, greater linguistic and ethnic diversity and more handicaps to learning.

After Sputnik, the U.S. instituted a massive campaign to identify and educate top science students. That effort was to project the U.S. into the technological lead once again. By the criteria of the race to the moon and the number of Nobel Prizes won, the U.S. achieved its objectives splendidly. Today the educational challenge is to inform and inspire the general student who will not necessarily become a scientist but who will have to go out and look for a job in an increasingly technical world and function as a responsible citizen. Can schools achieve this?

Peter Airasian, an education professor at Boston College, points out that schools have changed enormously over the past twenty years. They are bigger and costlier, have a more diverse student body and offer a wider range of programs and curricula. Schools are more politicized between boards, administrators and staff and between teachers in different areas. These factors make it more difficult to bring about change in the schools.

Airasian suggests that perhaps we should stop relying on schools to change social and institutional problems. But if greater science literacy is a goal, how can it best be achieved with the human, material and economic resources that are available?

Part of the reason for the decline in science performance of children in schools over the past three decades is the explosion in knowledge in scientific and technological fields. But there is a more fundamental problem. Teachers and students themselves have changed enormously since the Second World War. Teaching is no longer a calling, it is a profession, and the major rewards are external (raises, promotion, prizes) rather than internal (pride, satisfaction). Ambitious teachers feel compelled to get out of the classroom. In Los Angeles, for example, half of each year's new teachers have left teaching within five years.

Society has changed and the schools reflect it. The 3Rs now encompass a melange of offerings ranging from Rape to Retroviruses. Schools have to compete for a student's attention with numerous extracurricular activities, especially television. Divorce and single-parent families affect a significant proportion of the student population, who often require special counselling and support services.

Why should students take science? There is a twofold goal implicit in science education — recruiting students for science- and technology-related careers and providing basic science skills for civic responsibility in an increasingly technological world. But personal *attitudes* that affect whether people take more science are established long before children reach high school. Audrey Champagne, Program Director of the American Association for the Advancement of Science points out that young children quickly develop concepts of how the world operates. Many graduate from high school and university, she says, holding the same views they had in kindergarten!

I believe students will show greater interest in science when attention and effort are focussed on children much *earlier*. In a study of factors that influence children's performance in science, it is found that in addition to labs, number of science classes and teacher enthusiasm, *parents* are significant. This isn't surprising, but it is worth emphasizing — parental interest and enthusiasm for science are reflected in their children. It's more than just choosing our children's school and faithfully attending PTA meetings — we have to *do* things with our children and show an interest. I also believe that parents have to actively restrict their children's television viewing time. Television is a poor model for thinking — programs blare out information in disconnected fragments as the images move at a rapid pace.

By far the most effective way to interest youngsters in science is to focus on the critical years — kindergarten and the first years of primary school — when lifelong attitudes are set. Yet governments traditionally invest the least amount of attention and money in that part of a child's education. Teachers in the earliest grades should be given the highest priority and support, increases in salaries and most of the additional money committed for science teaching. More support material is needed, as well as field trips, visiting experts, films, books and equipment (and I *don't* mean computers). As a parent, I have encountered only enthusiastic, willing teachers — they just lack the support.

What should be taught at these early levels? The word *science* fills many primary school teachers with fear, so let's drop the word altogether. Let's think in terms of *exploration, discovery, connections.* We're not talking about providing children with experiments and facts they need for a career in science. We're trying to find ways to instill excitement and interest in the world around them.

Children are always excited when they learn about their own bodies and when they discover other life forms and find that they aren't that different from us. They are astonished to see the Milky Way through binoculars. One of my nephews has loved rocks ever since he was an infant and with his parents' encouragement, he is now a rock expert. His hobby will serve him well for the rest of his life because enthusiastic, interested children are a snap to teach at any level.

Children should have very good hand-held magnifying lenses, terraria and stethoscopes. The schoolyard should be dug up to grow vegetables and flowers. Classes need binoculars and telescopes and lots of nature books. Teachers need an information network to share ideas and projects with one another.

Children who are interested in the world around them are much more likely to follow that interest into careers in science.

WONDER REGAINED: WHAT EDUCATORS CAN DO

Introduction

*I*n order to prepare our children for a radically changed world in the next century, we have to re-examine the content and methodology of education from kindergarten through university. Too often, we are conveying completely unintended messages that shape the outlook of our children.

An Education Committee of the Science Council of Canada spent three years examining science education in elementary and high schools across the country. They described an inadequate preparation of youngsters for a world dominated by the products and consequences of science and technology. In order to correct this deficiency, the committee prescribed a series of steps that would lead to the goal of one hour of science instruction every schoolday of a child's life from kindergarten through high school.

Even if this is achieved, the content of the courses must be relevant to the lives of those children when they become adults. Often that isn't the case.

The Right Stuff

Years ago I read a marvellous book entitled *Is There Life After High School?* In spite of the title, it was a serious comparison of human relationships at different stages in life. The study revealed that impressions formed in high school are more vivid and indelible than those formed at any other time in life. The author described how people in their seventies and eighties who had difficulty remembering most of their associates in university and at work would instantly recall most of their classmates by name while leafing through their high school yearbooks. In the analysis of the author, high school society is divided into two broad categories, the innies and the outies. The innies were football and basketball players and cheerleaders who set the whole social climate of the school. The outies were all the rest, the majority of the student body, most of whom lusted to be innies. I sure hope it's different today because that description fits my recollection of high school and it was awful. But I'm getting off the point.

Those high school memories are so intense because that is the time when puberty occurs. The enormous physiological changes that take place in response to the surge of new hormones through the body completely transform both anatomy and mind. I always feel kids lose about half their intelligence for a few years in response to that blast of hormones. Relationships change radically. Suddenly parents change from protective, loving gods to dictatorial wardens incessantly imposing restrictions and criticizing everything. A pubescent teenager perceives adults and members of their own age group with totally new eyes. It's not surprising then that attitudes to school, courses and studying also change dramatically.

In the early 1970s, I visited a small northern town to judge a science fair. Back then, it was a tough town with a transient population of men working in the oil fields and a high proportion of native people. The night I arrived, I dropped in to the bar of the motel and a man came over and said, "I hear you're going to talk to the students at the high school tomorrow." When I affirmed it, he shocked me by adding, "They'll kill you. I'm the science teacher there and I can tell you that all they think about is sex, drugs and cars. They'll tear you apart."

Well, he really scared me. I immediately formed images of a blackboard jungle, filled with switchblades and drug-crazed hoods. The next day when I walked into that auditorium, it was with great trepidation. There were 400 teenagers in the gym, about a third of them Indians. They looked pretty normal, but I had been warned and knew they were just biding their time before turning into raving animals.

So I began by saying, "I'm a geneticist. I know that you're basically walking gonads, so I'm going to talk about sex." That opener caught their attention. I started with the beginning of human life by describing eggs and sperm, talked about chromosomes and the X and Y basis for sex determination and went on from there. The kids were dead silent and attentive. I talked for about an hour and then opened it up for questions. I was astounded at the range of topics we covered. We discussed drugs and chromosomes, test-tube babies, amniocentesis and cloning.The principal finally had to step in to dismiss the group an hour and a half after that.

Science education in high school should be designed around sex and human biology. It's a shock every time I hear that a school board has caved in to pressure and kept sex education out of schools. I am sure opponents of sex ed have no intention of providing that information to their own children. In a time of easy access to the most explicit films, videos, magazines and books, who can believe it's better to keep youngsters ignorant by denying them some accurate facts? They're going to get all kinds of anecdotal, apocryphal stuff about sex from their peer group, anyway.

By starting their instruction with human sexuality and reproduction, teachers will be able to go on to practically every other subject in science. It just takes a hard look from a different perspective. After all, we are not trying to train future scientists (only a small percentage of high school graduates will go on in science), yet all of them will be able to use information that science can provide for the rest of their lives. And you can bet they will remember those lessons vividly in their life after high school.

The System and the Ecosystem

*T*he fateful year of 1984 has already long passed and we're only about a decade away from the twenty-first century. What will the world be like then? I do not believe in futurism, the claim that the future can be predicted scientifically. We can extrapolate by project-ing trends into coming years, and it is possible to render events more likely by creating self-fulfilling prophesies. But that isn't prediction.

If the future is unknowable, the best strategy for our young people is flexibility and a broadly based education. Most of all, they should have an intimate connection with nature.

Any child who knows that a dragonfly can fly off before it can be grabbed, observes that a seed will germinate and send roots *down* and leaves *up*, watches a squirrel leap deftly from branch to branch, has envied birds soaring effortlessly and has followed a caterpillar through metamorphosis into a butterfly understands that no human technology can ever come close to matching living creatures. Our current techno-logical achievements are impressive by human standards, but mea-sured against the scale of life's complexity, they must be seen as crude and superficial. This perspective has to generate some humility.

A child with love of nature also recognizes that we *share* this planet and that we derive our sustenance from other life forms. If we forget that packaged eggs or hamburger came from animals, a cotton shirt from a plant, a wooden chair from a tree, then we have lost that connection with nature. In November 1985, I watched a performance that brought that reality back into focus.

I was at a meeting in London on the subject of toxic wastes. After my rather ponderous talk, Jack Vallentyne, an ecologist from the Centre for Inland Waters in Burlington, got up. Jack wanders the planet carrying a large globe of the Earth resting atop a backpack. It looks . . . well . . . unusual. He performs for children around the world by assum-ing the character of "Johnny Biosphere." Like Johnny Appleseed, Vallentyne plants the seeds of ideas in the fertile soil of children's minds. He put on quite a show, transforming an audience of some 700 self-conscious and skeptical adults into wide-eyed children who shouted answers back to his questions.

The most effective part of Jack's show was the way he demonstrated how much we are a part of the ecosystem. He asked all of us to hold our breath for a few seconds and then informed us that we had held gas molecules in our lungs that had been in the lungs of everyone else in the room. Initially, it made us want to stop breathing, but it brought home with a jolt the reality that we share the air with everyone else. Then Jack told us that we all have molecules in our bodies that had once been a part of every single human being who had ever lived in the past 3,000 years! (And he didn't even mention animals and plants.) From Jesus Christ to Marie Curie to Michael Jackson, we are all linked by shared molecules in the air, water and soil.

Jack then went on to tell us a story about an Indian canoeing on Lake Superior 500 years ago. It was a hot summer day and he was sweating, so he decided to go for a swim. And as that Indian swam in the lake, the sweat washed from his body and was diluted in the lake. Sodium and chloride ions in the sweat diffused through the lake and today, 500 years later, when we take a drink of water in Toronto, we drink sodium and chloride from that Indian's body long ago.

This is a modern description of a spiritual vision of our relationship with all life on earth. It is fitting that his target is children. The difference between Vallentyne's approach and mine is worth noting. I tend to emphasize the destructiveness of a value system that sees humans outside the ecosystem and all of nature as a potential resource. I believe we must recognize that the limited vision of science and technology does not give us control, so then we may try to change directions. My operating faith is in the power of reason to overcome cultural values that are generations old. But it's a bleak picture. Vallentyne uses a radically different approach — his message is just as dark, but it's delivered to children who have not yet accepted all our cultural values. He can revel in the unity of all life in the biosphere in a spiritual way that is both uplifting and wonderful. And I think he's on to something.

It Starts in Kindergarten

*I*n the summer of 1988, I received a remarkable letter from a kindergarten teacher in Hamilton, Ontario. I present it here almost in its entirety because it offers a profound message in a simple and moving way. The letter is from Catherine Verrall and was addressed to the Hamilton Board of Education.

"Look, teacher! See what I found!" I remember when a Junior Kindergarten child would call me to come see a ladybug or a tiny flower — surprise, a free gift — on our playground. But not anymore. The Hamilton Board of Education seems to have decreed that all nongrass plant life shall be killed — by spray (MCPA) or salt. So all along the fences, we have eight to eighteen inches of alkaline, dead soil, still sparkling with salt, but barren of all the interesting plant and insect beings who used to live there.

Remember purslane, luscious patches, chock-full of vitamin A and iron, delicious in salads and soup, prized as a vegetable in Asia and Europe. And common mallow, with its lovely little flowers and its edible "fairy cheeses" (fruit). There used to be clover (count the leaves by threes, search for a four-leaf). And Creeping Charlie, with its gingery smell and tiny blue flower. And dandelions to delight in, to pick and count, to feast a guinea pig. And there might be wild morning glory, gaily climbing the fence. There used to be tough-rooted plants (i.e., "weeds") holding the soil, absorbing the moisture, cushioning falls, on busy spots such as under the climber. But they too have been killed, so in their place we have patches of mud (in melting or rainy weather) and patches of loose dust in dry weather. Result: for two months in spring, we could not use the playground because of the mud; and on dry days, the children are coated with dust. Grass, a delicate hybrid, can't withstand the wear on these spots, but "weeds" could, if allowed to do their work. And when the drought comes, only the surviving "weeds" are green. Grass needs extra water.

For "Environmental Studies" we are supposed to notice differences in leaf shapes. We are advised to buy exotic foreign plants from a shop. Why not notice the even greater variety in the natural plants freely given in our everyday environment (unless they are banished)? We are supposed to teach about "Critters," yet we destroy the natural habitat of insect creatures — ladybugs and butterflies aren't much interested in short-cropped grass. What is there for the birds to eat? We are advised to make butterflies from tissue paper, caterpillars from egg boxes — substitutes for the real things.

The schoolground should be recognized as an integral part of the educational environment (actually, part of the "real world" from which the classroom is only an abstraction). The Ministry of Education document *Science in Primary and Junior Education: A Statement of Direction* (1986) declares that we must facilitate children's interaction with their environment including "the natural exploratory activities of children's play" because "from their youngest years, children are curious to find out about their world; they explore, they inquire, they ask questions, and they enjoy the process of discovery." On the schoolgrounds of Hamilton?

[T]he absence of interesting life trains children *not* to look, *not* to notice . . . .

Beyond the educational deprivation in this policy of "non-grass genocide," we need to look at the basic values being taught. When children came out one day to find every plant (except grass) lying brown and dead or piled over with salt, they asked "What happened?"

Teacher: Someone killed them.

C:Why?

T:They didn't want the plants to be here.

[W]hat are they learning? Nature is not a nurturer, our first Mother, but an Enemy. The gifts of Nature are welcomed only if humans control them. The words inside the school (and inside the Ministry's science documents) about "respect for living beings" are not believed outside. Technology kills. It's all right to kill for no good reason. Since Nature is an Enemy, people can be enemies. System-organized violence against Nature is okay on the schoolground, so what's wrong with violence between *people* on the schoolground?

Maybe we adults don't see these connections, but the children are more perceptive than we think. And they are growing up in a fearful world. They desperately need to know that we truly are at home in the universe, nurtured by Creation.

For our very survival, we all need to return to respect and caring for our natural environment. The attitude that says it's okay to kill "weeds" because we don't like them (disregarding the consequences) is not far different from the attitude that says it's okay to burn the rain forest for short-term advantage, disregarding the suicidal effects on the whole earth . . . . Herbicides, even if "Ministry approved," add more killing chemicals to our lives (e.g., draining off into the lake for us to drink). Should I let the children frolic in newly cut grass that has just been sprayed? Or grow lettuce beside a sprayed lawn?

Caring and not caring are learned in kindergarten. Please help us.

Verrall's letter is not only a powerful argument against using chemicals in any schoolyard, but it also penetrates to the deepest questions of what education is and how we are inadvertently passing on destructive ideas.

A Gifted Teacher in Japan

*T*he Japanese word *sensei* means *teacher* and implies a great deal of respect, because teaching is such an honoured profession in Japan. In March 1987, I attended a meeting in Yokohama, where I met a remarkable *sensei*, Toshiko Toriyama. She is a lone voice fighting against a rigid educational system that she says is taking a terrible toll among Japan's children.

Children are put under incredible pressure to perform, beginning in kindergarten and continuing right through high school. That's because their future opportunities depend so much on which university accepts them. Toriyama told me the system is so rigid that by the third week in grade 1 a child must have reached a specified page in the textbook.

Toriyama has rebelled against this system and tried a radically different approach that has got her into a lot of trouble with the educational authorities. They have tried to transfer her to teach the upper years in primary school, where the children are less impressionable. She has fought to stay with grades 1 to 3, where she believes she has the greatest effect. She may have been turfed out by now.

Japanese school schedules are very rigid; classes begin on the hour, and each subject is taught for fifty minutes. Toriyama designed a flexible schedule, letting the children pursue a topic for two hours or more, depending on their interest and concentration. Toriyama was galvanized into action when she realized that her students, who live in the heart of Tokyo, had never touched a live animal and often did not realize that neatly packaged food in stores was once living organisms.

At the meeting in Yokohama, she demonstrated her teaching techniques before about 500 people. She began by asking questions such as, "Were you alive before you were born?" Then she would answer that each of us lived inside our mother's body for about 270 days, starting from a fertilized egg. "What other animals begin their lives as eggs?" she asked. Gradually we realized that *all* other animals start out the same way. She then took us through growing up and getting old and dying. "All people die; it is nature's way," she told us.

Toriyama then went on to show us the life of a swallowtail butterfly from the time the eggs are laid on a pepper leaf, through hatching, moulting and pupation. She showed pictures of newly emerged butterflies and of butterfly predators such as spiders and rodents. Then she showed how the butterflies mate, lay eggs and die. She did the same for praying mantises and coral organisms. She had us close our eyes and imagine we were insects and soon we were wriggling to get out of our skins and dodging animals trying to eat us. It was a brilliant way to show our similarity to other organisms — to demonstrate our relatedness with other life forms.

Toriyama then described an electrifying experiment. She took a grade 3 class on a trip to the country, something almost unheard of in Japan. Once there, she told the city kids to go off and play on their own. In a couple of hours, they began to straggle back, tired and hungry. As they all gathered round asking to be fed, she produced a live chicken.

They were fascinated with it as she showed its feathers, claws and so on. Then she announced, "This is your meal." The children were horrified. Some cried and others begged her not to hurt the chicken. So they had a talk about food, and for the first time they realized that the cooked meat they ate had once been a living animal.

Eventually, hunger overcame their reluctance and the children agreed to eat the chicken. Toriyama helped them kill the bird, pluck it, clean it and eventually cut it up and cook it. It was a profound lesson, but the school authorities were furious with her. However, the parents were astounded to see a remarkable change in their children's attitude as a result of that experience — they were deeply aware that they depended on living things for their own nourishment.

Toriyama took a live pig to class, again educating the children about the animal. Then she took them to a slaughterhouse where pigs are slaughtered and she had the children hold parts of the viscera — the heart, lungs and stomach.

"What food do we eat that did not come from a living organism?" she asked us. I was amazed to reflect on it and could only come up with two things — salt and water.

Toriyama's radical approach is much more than a way to educate children about the facts of life. It is an attempt to reconnect these children to their biological roots. To her, nourishment, sexuality and death are natural, to be accepted and appreciated. I'm sure her students will come out as very different adults from graduates of conventional classes. Toshiko Toriyama is a tiny voice amidst the cacophony of high-tech barkers. She will not last long in a society determined to prepare children for the computerized world of the twenty-first century.

A Love Story

*B*ack in 1978, in his home village of Old Lyme, Connecticut, world-renowned bird expert Roger Tory Peterson was experiencing a complaint most often thought of as an affliction of artists and writers — creative block. He had to paint a picture, but try as he would, he just couldn't get a handle on a concept for the painting. For agonizing months, it stymied him.

His wife, Ginny, suffered along with him and desperately tried to think of ways to get him over the hurdle. How about something to do with birds, she wondered. Obviously birds were his great passion, but their home was filled with bird mementoes and surrounded by feeders and birdhouses. Clearly anything birdy would be a bit redundant.

Ginny then recalled that on their numerous field trips, Roger would become just as excited by *butterflies* as he was over birds. He'd follow, watch and photograph them. Could something be done to please her husband with these lovely insects?

Then suddenly one day in a moment of inspiration like Archimedes leaping from his tub yelling "Eureka," Ginny thought "Why not build a butterfly *garden*?" It was a brilliant idea — she'd establish a garden of plants that would attract colourful lepidoptera from the wild to flit about and delight Roger with their delicate aerial gymnastics. She decided to build it outside a window he passed every morning on the way to the bathroom.

But the idea was the easy part. Making it happen was a lot tougher. She couldn't find anything on butterfly gardens in the library. Then she found the Xerxes Society, an organization dedicated to preserving butterflies. They gave her literature and references. She discovered that there already were butterfly gardens in England, but much of the information in their books didn't apply to Connecticut. So she did what any good scholar would do — she spent months in libraries studying everything known about butterflies. She learned about their life cycles, their behaviour and their food plants. She became an expert.

In the spring of '79, she started with a small plot about six metres by seven metres. She knew that in order to attract the adults she would need wild flowers, many of which grew in marginal soil of the kind found along railroad tracks. So she began to walk along tracks hunting for treasured flowers that might tempt a specific butterfly into the Petersons' yard. More than once she was challenged by railroad inspectors who found it hard to believe she was really only looking for plants for a butterfly garden!

Ginny decided early on that she wouldn't just have flowering plants with nectar for adult butterflies. She wanted to have the insects living in her garden for their entire life cycles. That meant she had to have plants that the caterpillars ate and in numbers that would support them. Each year, her efforts and the garden expanded, and long after Roger had got over his painter's block, the garden had become a major project. However, about three years after Ginny had started the butterfly garden, a man from the heating oil company came to fill up their tank. He dragged the hose through the yard and accidentally spilled fuel all over the garden. It was a disaster for the plants and the butterflies.

By now the garden had become an obsession, and the next spring, Ginny dug out all the contaminated soil to a depth of about twenty-five centimetres, replaced it with fresh soil and started all over again. Today the butterfly garden flourishes and has grown to a plot that is thirty-five metres square. It is full of adult and larval food plants that last year attracted thirty different butterfly species! She tells me she's close to saturation at about thirty-three species.

But this love story doesn't end here. Word of the butterfly garden seeped out, and Ginny has been the subject of a number of national television programs. She notes that our obsession with having everything neat and tidy means that most states have programs of spraying and mowing along highways to keep weeds down. But plants are weeds only in human terms and many of them are flowering plants that nourish butterflies. Ginny has been outspoken in suggesting that if roadside cutting must take place, at least it ought to be delayed until the butterflies have been able to take advantage of the nectar in the flowers. Some states are taking her advice.

This is a wonderful story for a number of reasons. Mrs. Peterson says that a lot of people want to become involved in some way with conservation, but don't know what they as individuals can do. Butterfly gardens offer an immediate way to get involved. She has a circle of women who visit to share information and advice for their butterfly gardens. And as these new wave gardeners learn about butterflies, they may discover a new esthetic based on beauty in "wild" nature that has so much more complexity, interaction and unpredictability than our manicured lawns, gardens and highways. Why not have butterfly gardens in every schoolyard in the country?

We Are Part of the Web

Science is a uniquely human activity, a way of learning and knowing that comes from focussing on parts of nature. By separating a fragment from everything else, controlling everything impinging on it and measuring everything in it, we gain powerful insights — on that isolated *piece* of nature. What we accumulate is a mosaic of disconnected fragments that do not add up to a complete picture. Our insights into each bit may be profound but of little value in predicting how different bits will behave when they interact. There are secondary, tertiary and quaternary levels of reaction and interplay that the limited view of science cannot comprehend. This disjointed perspective also prevents us from anticipating the full consequences of our application of new technologies.

So is science destined to dig out only discontinuous bits and pieces? Not necessarily. For emerging out of science itself is a new vision that *connects* us with everything in the universe and attaches us in the web of life on earth. This scientific version of our origins is a rephrasing of what are often thought of as *primitive* worldviews.

In trying to understand the beginnings of the universe and the basis for its expansion, astrophysicists now theorize that matter as we know it did not exist some fifteen billion years ago. It was all compacted within an unimaginably dense mass that exploded with a Big Bang, hurling matter outward in a universe that continues to expand today. In those early moments of the universe, matter began to coalesce into more familiar states and eventually formed gaseous clouds that condensed into stars and planets. We, and any other life forms that may exist elsewhere in the universe, are made of the atoms that were cooked in the cauldron of the Big Bang. As Carl Sagan says, "We are made of star stuff."

Ecologists now look at the skin of life on this planet as a collage of interconnected life forms that *share* water, food and energy. All our nutrition comes from other living creatures. Our bodies contain atoms and molecules that once resided not only in every other human being who lived over the past thousands of years, but in other mammals, birds, trees and insects. The very stuff we are made of is an accumulation from the past and compost for future generations.

And *evolution* has provided the cohesive explanation of the rich diversity of life on the planet and an idea of our place in the scheme of things. After the planet formed some 4.7 billion years ago, life appeared within a billion years. From the rich soup of molecules that was formed and accumulated over hundreds of millions of years, prototypes of life came into being and died out.

One of them managed to survive, reproduce and spread. That cell was one of many, but it was unique because all other life forms that came to occupy the planet were its descendants. And ever since, all organisms have been built on the same basic plan. The main atoms composing all life are the same — carbon, oxygen, hydrogen, sulphur, nitrogen, phosphorus. The building blocks of large molecules are identical in all organisms. And all life is based on the same basic structure — *cells*. Every creature alive today can trace his, her or its ancestry back to that first primordial cell. All life on earth is truly related. When native people speak of the tree people, their raven and whale brothers and sisters, or their wolf and bear clans, they speak in more than metaphors — science informs us that there is a physical basis for it.

Within the cells of complex organisms, it appears that cellular "organelles" (complex structures within cells) were originally free-living organisms that parasitized other cells. These invading cells — bacterialike organisms — came to coexist in their hosts. They derived all the necessities for reproduction and growth from their hosts while in turn performing functions that ultimately benefitted the infected host. Eventually, the "symbiotic" organisms became so intimately integrated into the host that neither could exist any longer on their own. Each of us, then, is an expression of cooperation — we are literally made possible by a myriad of other creatures that are integrated into all our cells. We know from molecular biology that all life forms are organized according to specifications contained in nucleic acids, DNA or RNA. And when the information carried within those molecular blueprints is compared, the results are astonishing. Comparison of human DNA with DNA of chimpanzees and gorillas reveals that ninety-nine per cent of our genetic information is *identical*. We are as closely related to the great apes as they are to one another.

And while there is a great deal of variation from individual to individual human being, average differences *between* races are less than the range of variation *within* a race. Genetically then, the differences between all people are truly only skin deep. And now a study of DNA contained within cell structures that are passed from generation to generation primarily through females suggests that people in all races can trace their ancestry back to one female in Africa about 200,000 years ago! There is a brother- and sisterhood of all humankind.

Thus although science can only function by focussing on parts of nature, the insights we are gaining provide us with a picture that *connects* us with each other, with all life on the planet and beyond to the rest of the universe. It's at once humbling and yet comforting, an idea of our place within all of nature.

THE ABORIGINAL WORLDVIEW

Introduction

*T*he rapid increase in human numbers at the very time that technological muscle power has also leapt ahead has generated the environmental crisis today. Scientists like Harvard's Edward Wilson and Stanford's Paul Ehrlich point to the need for a profound shift in attitude towards the natural world. The change must be a "quasi-religious" shift in the spiritual value that we place on other organisms.

Many of us sense in our guts that something is drastically wrong with our obsession with consumption and profit. But is there an alternative? Surprisingly, right here amongst us is a worldview that offers a profoundly different vision of the human place in nature. If we paid attention to it, we could restructure our priorities and return to a balance with the environment. Ironically, this worldview is found among the most oppressed and deprived groups in our society, the aboriginal people.

Children of the Earth

Nothing is more moving than the authentic record of the testimony of Chief Seattle in 1854. He gave a magnificent speech to an assembly of tribes who were preparing to sign a treaty with the white man. He delivered his oration in his language, Duwamish, and it was recorded by a Dr. Smith who maintained afterwards that his translation did not do justice to the chief's imagery and thought. (The entire speech is printed in *Fellowship*, vol. 42, no. 11, from The Fellowship of Reconciliation, 523 N. Broadway, Nyack, N.Y. 10960). This is wisdom we could learn from today:

How can you buy or sell the sky, the warmth of the land? The idea is strange to us. If we do not own the freshness of the air and the sparkle of the water, how can you buy them? Every part of this earth is sacred to my people. Every shining pine needle, every sandy shore, every mist in the dark woods, every clearing and humming insect is holy in the memory and experience of my people. The sap which courses through the trees carries the memories of the red man.

The white man's dead forget the country of their birth when they go to walk among the stars. Our dead never forget this beautiful earth, for it is the mother of the red man. We are part of the earth and it is part of us. The perfumed flowers are our sisters; the deer, the horse, the great eagle, these are our brothers. The rocky crests, the juices of the meadows, the body heat of the pony, and man — all belong to the same family . . . .

The ashes of our fathers are sacred. Their graves are holy ground, and so these hills, these trees, this portion of earth is consecrated to us. We know that the white man does not understand our ways. One portion of land is the same to him as the next, for he is a stranger who comes in the night and takes from the land whatever he needs. The earth is not his brother, but his enemy, and when he has conquered it, he moves on. He leaves his fathers' graves behind, and he does not care. He kidnaps the earth from his children. He does not care. His fathers' graves and his children's birthright are forgotten. He treats his mother, the earth, and his brother, the sky, as things to be bought, plundered, sold like sheep or bright beads. His appetite will devour the earth and leave behind only a desert . . . .

I am a savage and I do not understand any other way. I have seen a thousand rotting buffaloes on the prairie, left by the white man who shot them from a passing train. I am a savage and I do not understand how the smoking iron horse can be more important than the buffalo that we kill only to stay alive . . . .

What is man without the beasts? If all the beasts were gone, men would die from great loneliness of spirit. For whatever happens to the beasts, soon happens to man. All things are interconnected . . . .

You must teach your children that the ground beneath their feet is the ashes of our grandfathers. So that they will respect the land, tell your children that the earth is rich with the lives of our kin. Teach your children what we have taught our children, that the earth is our mother. Whatever befalls the earth befalls the sons of the earth. If men spit upon the ground, they spit upon themselves . . . .

Where is the thicket? Gone. Where is the eagle? Gone. And what is it to say goodbye to the swift pony and the hunt? The end of living and the beginning of survival.

This we know. The earth does not belong to man; man belongs to the earth. This we know. All things are connected like the blood which unites one family. All things are connected.

Whatever befalls the earth befalls the sons of the earth. Man did not weave the web of life, he is merely a strand in it. Whatever he does to the web, he does to himself . . . .

The whites too shall pass; perhaps sooner than all other tribes. Continue to contaminate your bed, and you will one night suffocate in your own waste . . . .

When the last red man has vanished from this earth, and his memory is only the shadow of a cloud moving across the prairie, these shores and forests will still hold the spirits of my people. For they love this earth as the newborn loves its mother's heartbeat. So if we sell you our land, love it as we've loved it. Care for it as we've cared for it. Hold in your mind the memory of the land as it is when you take it. And with all your strength, with all your mind, with all your heart, preserve it for your children and love it.

This is not a romantic fiction but an authentic transcript. Nor is it a record of an extinct civilization. The deep relationship with nature and the land so evident in Chief Seattle's eloquent plea can be seen and heard over and over again from pockets of native peoples throughout North America. All we have to do is listen.

In 1983 and 1984, Thomas Berger travelled across Alaska to listen to Inuit and Indians. He recorded some of their words in the book *Village Journey*, and they attest to the continued survival of the spirit of Chief Seattle:

> The land we hold in trust is our wealth. It is the only wealth we could possibly pass on to our children . . . . Without our homelands, we become true paupers.   — Antoinette Helmer

> We draw our identity as a people from our relationship to the land and to the sea and to the resources. This is a spiritual relationship, a sacred relationship. It is in danger because from a corporate standpoint, if we are to pursue profit and growth — and this is why profit organizations exist — we would have to assume a position of control over the land and the resources and exploit these resources to achieve economic gain. This is in conflict with our traditional relationship to the land. We were stewards, we were caretakers and where we had respect for the resources, that sustained us.   — Mary Miller

> The White men try to collect money and put the money in the bank, and whatever is put in the bank belongs only to them. But the Natives of Alaska used the land only for survival . . . . The bank was for everyone, not just one person.   — Jens Flynn

In 1985, former B.C. premier Bill Bennett established the Wilderness Advisory Committee to make recommendations on the future of several wilderness areas in the province. One of them was the Stein Valley, home for native people for thousands of years and still unlogged. John McCandless and Napoleon Kruger made a submission to the committee on behalf of the Lillooet Tribal Council, entitled "Stein Spirituality":

> The spiritual and cultural entities which are alive in wilderness are

"*priceless*," thus defying economic valuation . . .

In truth, ancestors still sing in the Stein, using words which have remained unchanged throughout centuries. In truth, there is laughter on the breeze and each tree, each stone is alive. That the ancestors have always protected the Stein, and always will, is true; and the Stein in return protects all. As various committees attempt to weigh trees against fish, and fish against deer, and deer against man against tree again, and *all* against money, it is laughter sweeping through the ages down the canyons which can be heard.

No archeologic site within Stein stands alone, no more than does any tree or stone or pool . . . . The long-standing tree people with their lofty spirits, the cool swimmers, the soaring flyers with their unbridled spirits: all contribute to the whole intact entity, wilderness Stein, whose spirit, in water, flows on to us all.

To the initiated, the Stein fairly bristles with spiritual power which would be lost with the construction of the first kilometer of road into this wilderness. The Stein basin is, in a way, a vessel which contains limited awareness, knowledge and power. The proposed road is the leak which could drain this watershed, if allowed.

Contrast those words with the submission from the Share the Stein Constituency Group, made up of politicians, loggers, lumber representatives, truckers and others who want to exploit the Stein for economic gain.

The case for development is simple. Is it possible in this great resource-based province to continue allocating huge tracts of forest land without paying a severe economic penalty?

Boston Bar, Hope, Lytton and Lillooet are four Fraser River communities whose futures are in peril if development in the Stein is not allowed.

Well, there it is. One view sees the Stein as the source of a spiritual wisdom and connection with nature that is beyond price, and the other regards preservation as a waste, a sin against the necessity to exploit for short-term economic gain. When the forest is gone, native people lose their spiritual centre and identity. When the forests are "harvested," loggers, truckers and the mills move on. If those who want to "share the Stein" really do care about survival of their communities, then they have an even greater incentive to preserve the forests. They have to make credible their claims that they can sustain the yields through reforestation. They have to work on diversifying the opportunities in the area and to add far more value to timber after it is cut down. Native priorities of spiritual values could be one of the greatest gifts to a society that seems hellbent on destroying everything now.

Showdown in Brazil

North American native people have undergone a remarkable transition: they were forced to yield large parts of their culture and traditions to take up the language and ways of the European invaders. While they have retained vestiges of the past, in the most remote parts of the world there are pockets of people who continue to live as nomadic hunter-gatherers the way all human beings did for ninety-nine per cent of our collective existence. Groups of these people are now being squeezed by forces of development that are tearing at their land. In Brazil, the greatest remaining ecosystem, the Amazon rain forest, is home to Indians who are determined to protect their land and retain their traditions. They are twentieth-century people who use all the modern technologies they need to aid their fight.

If you read newspapers very carefully, you may have noted that in mid-October 1988 South American Indians were blocked from entering a Brazilian court because they were improperly dressed. It is worth looking into the story hinted at by those brief news reports to understand the current plight of Brazil's indigenous people.

The Kaiapo Indians live in the Amazon rain forest much as they have for thousands of years. The Kaiapo are warriors whose primary weapons are the fearsome sight of their warpaint, bows and arrows and a deadly club they call a bat. Like aboriginal people around the world, they are under increasing pressure from people they refer to as the "civilized" ones.

Brazilian Indians are wards of the state and officially treated as "minors" by FUNAI, the government Indian agency. Native culture is not valued in Brazil, so FUNAI's goal is to assimilate the Indians as quickly as possible. Of course, assimilation also frees Indian land for other uses. Mining, farming, logging and hydroelectric dams have polluted the rivers, cleared the forest and squeezed the Kaiapo into an ever-shrinking area. Extinction threatens Brazilian Indians as well as the Amazonian flora and fauna.

The Brazilian constitution guarantees protection of Indian land, yet various intruders have steadily cut into Kaiapo territory that had not yet been demarcated. In 1985, angered by the mining going on in their land, a young chief named Paiakon led a force of 125 warriors on an eight-hour hike to the mine site. There they discovered a massive operation with 5,000 miners protected by armed guards.

Under cover of night, Paiakon and his tiny group seized the airstrip with seven planes, overwhelmed the sentries and their weapons and then captured the entire mining force! The Indians expelled the miners and held the camp and captive officials until three months later the government recognized the reserve boundaries, put the Indians in charge of the mine site and gave them a five-per-cent royalty on all gold taken from their land. It was an incredible victory achieved without the loss of a single life.

In January 1987, Darrel Posey, an ethnobiologist who has worked with the Kaiapo for over eleven years, accompanied Paiakon and chief Kube-i to a conference on tropical rain forests held in Miami, Florida. Through Posey, the two Indians spoke about the forest and rivers, what they mean to their way of life and the terrible effects of massive development projects.

The chiefs were especially concerned about the potential effects of a series of dams being planned for on the Xingu River. Costing $10.6 billion (U.S.), the dams will flood 7.6 million hectares, eighty-five per cent of it Indian land. The World Bank was preparing to give a $500 million loan to Electrobras, Brazil's power company to finance the dams. In Miami, the Indians were urged to go to Washington to talk to politicians and representatives of the World Bank. The chiefs did and had a widely publicized meeting with politicians and bank officials.

Upon returning to Brazil, Posey and the Indians were arrested for "criticizing Brazilian Indian policy" and "denigrating the country's image abroad." All three were charged under a law that forbids *aliens* from getting involved in issues of Brazilian national interest. Thus, indigenous people of the Amazon were legally classified as *foreigners!* Their first court appearance was scheduled in the town of Belem on October 14, 1988, and I was fortunately present to observe it.

That morning, soldiers arrived in front of the courthouse armed with shields, guns, rifles, clubs and bulletproof vests. Posey came early with his lawyer and they went into the courthouse. About eight-thirty, buses pulled up and disgorged about 250 Indian warriors fully dressed in warpaint and spectacular feather headdresses and armed with clubs and bows and arrows. Paiakon and Kube-i came in a car as the warriors filled the street.

The Indians began to sing and dance in front of the phalanx of soldiers in a powerful demonstration that evoked for me the emotional impact of drumming and chanting by Canadian Indians. One of the "civilized" spectators in the court muttered contemptuously, "How quaint." When Kube-i attempted to enter the court, he was stopped because, the judge said, he was "seminude." The judge ruled that Indians must dress to show respect for Brazilian law. When Kube-i replied that he was wearing his traditional attire, the judge answered that he had to follow Brazilian formalities and should strive to become a Brazilian. As Posey said to me later, the judge's policy was tantamount to genocide.

The Indians waited ten minutes for the judge to change his ruling. When he didn't, they departed peacefully, leaving word that the court could contact them in their forest homes. Suddenly and quietly, they were gone. The charges were eventually dropped, as much from embarrassment from the glare of international publicity as from any consideration of the issue.

Brazilian Indians are a repository of thousands of years of accumulated folk knowledge about the forest that can never be replaced. They are now fighting a desperate battle to save the remaining bits of their home. Their fight is ours because that great biological jewel is a part of the world's vanishing natural treasures.

I was privileged to meet Paiakon while filming in Brazil in September 1988. He asked me to help him raise money for his struggle, specifically for a major event he was planning for the following February. I called my wife, Tara, from Brazil and asked her to arrange for fund-raising events in November. She organized events in Toronto and Ottawa, and to our astonishment, Canadians responded massively — thousands turned out to hear Paiakon speak and tens of thousands of dollars were contributed for his cause.

That money was two-thirds of the amount that was used to bring several hundred Indians from the rain forest along the Xingu River to Altamire, the site of a proposed dam. In a huge media event, the Indians were joined by environmentalists, union leaders and civil rights supporters to tell the Brazilian authorities that the dam should not be built. Thirty-eight Canadians, including four natives, attended the Altamire meeting and were the largest contingent from any country. The event drew together Brazilian and international supporters who represented the nucleus of a burgeoning environmental awareness around the world that if Indians save their homes, wilderness will be preserved for all of us.

The Invisible Civilization

*I*n order to learn from native people about our relationship with the natural world, we have to recognize that they exist. To most of us, they are invisible.

A radio reporter is interviewing one of the early settlers in the northern part of British Columbia. The old-timer recalls that when he arrived in the area, "There was no one else around," adding as an aside, "just a *few Indians*." Now I'm sure the man meant to say that there were no other people of European background in the area, but his remark recalls Ralph Ellison's classic novel *Invisible Man*. Ellison's book was a searing commentary on the consequences of being black in the U.S., the most dehumanizing aspect of which was being invisible to the white majority.

On May 5, 1988, B.C. Minister of Forests Dave Parker met with a group of Indians from the interior of B.C. to continue negotiations on the future of the Stein Valley. Chief Ruby Dunstan of the Lytton Indian Band listed a number of grievances where Indian requests were repeatedly ignored or slighted. She ended with the plea, "Stop treating us as if we are *invisible*! We're human beings too."

Parker's reaction was astonishing. He took umbrage and huffed, "I deeply resent what you've just said. There's reverse racism too, you know." Chief Dunstan's plea to be treated with dignity was grotesquely twisted into a ministerial insult.

As in Ellison's novel, native people in Canada don't seem to exist except as government statistics. When Europeans arrived in North America, the continent was already fully occupied by aboriginal peoples with rich and diverse cultures. Yet we in Canada refer to the French and English as the "two *founding* races." This denial of the very existence of already thriving native people is reinforced by the cursory description of natives in our history books. Our governments have systematically oppressed and exploited native people, destroyed their culture and denied or opposed their right to claim aboriginal title.

The history of native people after contact with Europeans is a tragedy that continues to the present. Hunting and gathering nomads were forced to give up their traditional way of life for permanent settlements. Children were exiled from their remote homes to be educated in urban centres hundreds or even thousands of kilometres away. There they were taught to be ashamed of their culture and were punished for speaking in their own language. And we outlawed some of their most important, even sacred, cultural activities — the potlatch, the drums, dancing, religion, native medicines, even hunting and gathering.

Their original population ravaged by disease, dispossessed, forced to leave the bush and to abandon ancient traditions, Indians today occupy the lowest rung in social standing in Canada. It is hardly surprising that alcoholism, unemployment, suicide, sexual abuse and crime afflict many native communities across Canada. And these negative images are only reinforced by their constant repetition in reports by the electronic and print media.

I have recently watched three films about Canadian native people that had been commissioned by Indians themselves. What a difference they represent from the usual stereotypes in the media. The films allow Indians to talk about themselves and their own ideas, obstacles and goals. The reaction of media people who screened the films for possible television broadcast is instructive. "Beautiful footage, but too onesided" was a standard response. When a native in one film talked about the urgency of conserving salmon because the fish are at the core of his culture, a television executive snorted, "Are they trying to tell us they know something about ecology?" Another remarked, "If you leave it to the Indians, there won't be any salmon left." Another person suggested, "The women are too fat and the men look too white." When the narrator said, "Along these banks are remains of ancient civilizations dating back nine thousand years," one man interjected, "I don't think you should call them civilized because they were nomadic."

Underlying these responses is the assumption that media people know what reality is and how Indians should be portrayed. But there is no such a thing as objective or balanced reporting in the media. Everyone in the world is moulded by heredity, personal experience and cultural milieu, factors that shape not only our values and beliefs, but the very way we *perceive* the world around us. It's easy to confirm this — just talk to an Iranian and Iraqi, a Northern Irish Catholic and Protestant, a South African white and black or an Israeli and Palestinian about events that both in each pair have witnessed.

The point is that we edit our experiences through the lenses of our personal worldviews. That creation of subjective reality doesn't suddenly stop when people become part of the media. Human beings cannot help but impose their personal priorities, perceptions and biases on their reporting because *that is all they know*. Implicit in our newscasts are all kinds of biases — we just don't recognize them as such because they happen to be the dominant view in society. But ask a Soviet or Chinese visitor and they'll see the biases immediately.

As long as native people are not a part of the media, they will be portrayed as the rest of society chooses to perceive them. At the very least, the media ought to be willing to let natives present reports from their point of view, if only to render them a little less invisible.

Journalists often do hold up objectivity as a high goal in reporting. A few years ago, one of *The Globe and Mail*'s eminent reporters chastised me for publicly supporting a political party, because, in his opinion, it compromised my objectivity as a broadcaster. When I signed a letter supporting an antinuclear petition, I was informed that some in the Canadian Broadcasting Corporation management felt I had lost credibility on nuclear issues.

If there really is such a thing as "objective journalism," then surely there should be no need to worry about greater representation of women and people from visible ethnic communities in the media. Nor would it matter that there is a preponderance of upper-middle-class white males doing the reporting, as long as they were "objective."

If most reporting is truly objective, then there is no legitimate claim for the necessity of Canadian print or electronic media. If reporters simply observe and transmit objective reality, then the source of news ought not to matter. Of course this is ludicrous. The reason we value the CBC, the National Film Board, Canadian magazines and newspapers is that they present perspectives from within this country's culture. None of us can escape the limitations of our heredity and personal and cultural experiences. There's no such thing as objectivity.

There is plenty of evidence to show that complete objectivity does not exist even in that most rational and objective of all activities, science. Harvard's great science popularizer, Stephen Jay Gould, has written a marvellous book entitled *The Mismeasure of Man*. In it, he documents the history of the scientific study of the human brain and shows how existing beliefs and attitudes affect not only the kinds of questions asked and experiments conducted but also the way the results are interpreted.

So, for example, when it was believed that brain size was correlated with intelligence, scientists obtained evidence that the brains of blacks were smaller than the brains of whites. Decades later, when Dr. Gould measured the cranial capacity of those very same skulls, he found there were no statistically significant differences. By then it was also clear that brain size alone is not an indicator of intelligence.

Similarly, Dr. Gould describes how scientists once believed that intelligence could be pinpointed to a certain part of the brain. And, sure enough, when comparisons were made, that part of the brain in women was found to be significantly smaller than it was in men. Years later, when it was known that that particular part of the brain had nothing to do with intelligence, a reexamination of the data revealed that the differences were not significant.

We acquire genetic and cultural "filters" through which we perceive the world around us. I was struck with the power of those filters in 1987 when I visited the Stein River Valley. I was flown by helicopter up into the valley with my host, a Lillooet Indian. He pointed out the burial grounds of his ancestors, the battle site between his people and the neighbouring tribes and their ancient hunting grounds. Our pilot told me that a week before he had taken a load of foresters over the same area and all they spoke of were the number of jobs, the years worth of logging and the enormous profits those trees represented. The foresters and the natives were looking at the very same place, but "saw" things that were worlds apart.

Now you see why native people must have a means of seeing themselves through the lenses of their own values and culture; otherwise, they will live only with the fabrications of non-native journalists.

But I have a far more selfish reason for supporting them. I believe that while North Americans search for spiritual values of Eastern religions and African cultures, we ignore an important perspective right here in our midst. Native people have a very different view of their place in nature, and in spite of the way they have been brutalized over the centuries, they have hung on to those differences. Through a native perspective, *we* can measure and reexamine our own assumptions and beliefs. It is only in having a contrasting view that we can truly recognize our strengths and deficiencies. To do that, we have to abandon the myth that there is some high form of objective reporting and acknowledge our inherent and inescapable biases.

What the Innu Know

*F*rench scientists Claude Lévi-Strauss and François Jacob have pointed out that the human brain has a need to "make sense" of the surrounding world. Throughout history, people have created "worldviews" that are comprehensive explanations of the cosmic forces impinging on our lives.

Science differs from traditional ways of knowing by zeroing in on a specific phenomenon or object and ignoring everything else. This approach is the strength of science, but also a flaw, because in focussing on the subject at hand, scientists often can't see the larger context that it functions in.

We know very little of the spectrum of organisms that live on our lands and even less about their basic biology. We know next to nothing about how different species or groups of organisms interact with each other and their environment. There is no comprehensive, up-to-date inventory of terrestrial and marine plants, animals and microorganisms in North America.

But there exists a vast source of information about the environment that is every bit as profound and far more useful than what is learned from basic science — it is the body of knowledge accumulated by indigenous people around the world and preserved by those who cling to traditional ways and the relationship with the land.

The knowledge accumulated by aboriginal inhabitants about their wilderness environment represents thousands of years of keen observation. This knowledge was used to obtain food, medicine and materials for survival. It had to be accurate. I was struck by the value and extent of indigenous knowledge on a visit with Innu Indians in Labrador. They are struggling to stop low-level training flights by NATO pilots that, I can attest, are terrifying and profoundly upsetting. I would be astounded if feeding, reproduction and migration of the wildlife in the area were not affected by this disturbance.

Both federal and provincial governments have refused to recognize Innu sovereignty over the land, so the Indians have been forced to try to prove that low-level flights are detrimental to people and to the animals and plants on which they subsist.

It is completely unfair to demand that the victims of technology provide "scientific proof" of the technology's adverse effects. For one thing, it takes expertise, money and years of work to accumulate the kind of basic information that could prove whether there are any detrimental effects from the flights.

It is clear that the Newfoundland government is committed to stimulating the economy by expanding the airfield and the training flights. This puts the Innu under pressure to demonstrate a negative impact even though government authorities have no basis for saying there are no harmful effects.

However, the Innu already have a body of knowledge that demonstrates such effects. In speaking to Innu elders, I was astonished to learn about their intricate knowledge of caribou. Their culture is centred on these mammals. The Innu carefully observe their biological condition. They learn a great deal from the animals they slaughter for food, such as the state of their digestive and reproductive organs. They can judge a caribou's health by the bone marrow. They assess the environment by the normal colour, size and potency of medicinal plants. Their knowledge base allows the current condition of animals and plants to be checked against the past.

Since the arrival of the NATO jets, the Innu know that animals and plants alike have been adversely affected. This is not primitive, anec-dotal information, it is accurate and profound knowledge. However, in our linear, reductionist approach, we relegate native knowledge to the pages of anthropological studies. Thus we ignore what is a priceless potential reservoir of knowledge that becomes all the more precious as vast tracts of wilderness are dispoiled or disappear forever.

Descrating the Cathedral

At the fourth annual Stein Festival near Lytton, B.C., close to 4,000 people gathered to celebrate the physical and spiritual values of the Stein Valley. The participants were also there to show their support to the Lytton and Mount Currie Indian Bands who are fighting to prevent logging in the Stein River watershed. During the three-day celebration, two young Indian men independently described what the Stein meant to them in strikingly similar terms. Each referred to the valley as a "church" or "cathedral" where they could go to find spiritual sustenance and restoration. "When I come out of that valley," one of them said, "I feel that all my troubles are gone. I feel cleansed and refreshed, the way I do when I come out of my sweat house."

The festival in 1988 was held on the grounds of the Christian church where for over half a century, native boys and girls, including Lytton Band Chief Ruby Dunstan, were forced to come and receive their education in the infamous residential schools. As we left the festival site, we passed a lovely church nestled on the hills between the decaying dormitories and sheds.

That church brought to mind the great Christian cathedrals of Europe, the Soviet Union and North America. Many are stunning expressions of art and architecture. The music of organs and choirs echoes through these magnificent edifices, adding to the elaborate paraphernalia of worship. In contrast to the Stein Valley, these temples are essentially celebrations of our species' creativity, originality and cleverness. Our cathedrals are an affirmation that *we* are special, we are God's chosen.

In contrast, in most native cultures, it is the *land* that is sacred and everything that is part of it — rivers, fish, trees, rocks — was put there by the Creator. Thus, to bulldoze away a sacred rock so a road can be built, to clearcut a watershed or to reroute a creek is tantamount to sacrilege.

In a human-oriented religion in which all of nature is seen as God's gift to us as the chosen species, an untouched watershed becomes an opportunity. "Go forth and multiply, and fill the earth and subdue it . . . And have dominion over the fish of the sea, the birds of the air and every living thing that creepeth upon the earth." This notion of our special relationship with God with the Earth as a storehouse for our use is the foundation of Western attitudes and values, even though we are no longer aware of it today.

A clever and inventive species, we are able to think up all kinds of things to do with what we see around us. In this century, we have achieved an unprecented level of health and material comfort. Our ingenuity now expressed through science and technology is taking us to places that were hitherto unreachable — outer space, the bottom of the oceans, deep into the earth. Within our Western perspective, the conception of a use for our surround is sufficient justification for acting on it.

Now that we have set our sights on colonizing outer space, engineering organisms for release into the wild, creating intelligent machines, constructing biological weapons and developing more elaborate machines of destruction, we cannot refrain from pressing on. We justify our actions with the rationalization that if we don't do it, someone else will.

Thus, as we gaze at the Stein Valley, we see it in terms of jobs, board feet and profit. We look to harvesting mineral nodules from the ocean depths, tapping oil and gas from deep beneath the Arctic ice, exporting fresh water that flows "uselessly" to sea. Once we have imagined a possibility for use of these "resources," to refrain from doing so is a waste, a sin against a generous God.

But the Judeo-Christian exhortation to subdue the earth is only one expression of our connection with nature. We needn't look to mystical religions in the Far East for inspiration because there are other passages in the Bible that speak of sharing, stewardship and belonging. Jesus Christ himself needed no impressive temple or elaborate robes to worship God. His temple was in the open on the mountainsides, the shore of Galilee or the desert. And one of the profound interpretations of the Western tradition was expressed by St. Francis of Assisi. St. Francis recognized the kinship of all life and our spiritual equality. He felt at peace in fraternity with nature and through that harmony worshipped God. Thinking about the significance of the Stein Valley and other wilderness areas like it suggests that therein lies the potential for a reinspection and reinterpretation of our Western roots.

The Spiritual Value of Land

Can those of us who live with the attitudes of Western societies recognize anything of native values in ourselves? I believe that there are intimations of noneconomic, intangible factors that matter deeply to us.

I received a form letter the other day from a real estate agent. It seems that Vancouver real estate is "hot" right now and I can make a lot of money by putting my house on the market. It is said that there is a lot of "offshore money" flooding into Canada for investment in property and that drives land values up. People are buying houses as a hedge against inflation. And it is precisely this way of treating land as mere property for economic transactions that is leading us into a deep ecological crisis.

I will never willingly sell my house because the things that make it home have no economic "value," yet for me are beyond price. You see, in the fifteen years that I've owned my place, it has accumulated a history that is very real for me. My best friend, Jim Murray, helped me build the backyard fence and carved a gate handle. They remind me of him every time I enter the yard. Inside the house we still use a kitchen cabinet that my father made for Tara and me when we were just married. We salvaged it from our apartment because it represents a bit of Dad and our early years together. One of our prized possessions is an old table that recalls the friends who gave it to us when they left Vancouver to return to Winnipeg. Each year I pick asparagus and raspberries that my father-in-law (who lives upstairs) planted for me because he knew how much I enjoy them. Our English flower garden is his pride and joy, and every time I admire it, I can picture him standing there, leaning on a shovel and puffing his pipe. Every Christmas brings memories of glorious feasts prepared effortlessly by Tara's mother.

We buried the family dog, Pasha, under the dogwood tree in his favourite digging hole. My daughters have turned that area into a cemetery for a hamster, a salamander and several other dead animals that they've found around the neighbourhood. In the branches of that tree is one of the girls' favourite play areas, a treehouse that I spent many happy hours building and many more just watching children playing in. And everywhere in the house and yard, my home is alive with reminders of my wife. A clematis plant has climbed along the back gate. Four years ago after my mother died, we scattered her ashes on the ground around it. Last year we added the ashes of my sister's daughter. Now when the purple flowers bloom, the pain of the loss of my mother and niece is softened because somehow I feel they are nearby.

In the real estate market, none of these things adds a cent to the "value" of the property, yet for me they are what make my house a home and represent value beyond price. In less than two decades, family and friends have become a palpable presence everywhere. What I'm talking about are things that exist only in my mind, memories and experiences that are an expression of my personal values and history. These are essentially spiritual values. I have accumulated memory and history in only fifteen years of occupation; native people have lived on their land for thousands of years.

Years ago I visited the Dome of the Rock in Jerusalem. That beautiful Muslim temple covers one of the most sacred places in all of Islam — a rock! A visitor from outer space would find completely incomprehensible the fuss made by a group of people over what is a rather ordinary geological formation. What makes it such an important site is the fact that the prophet, Mohammed, is said to have ascended to heaven from that rock. Human beings assign worth to objects, places or people on the basis of factors that have no physical or material reality — they are really in the mind.

Our species has a great capacity for seeing value in intangible things. But as people in the highly industrialized countries have become increasingly mobile, our sense of permanence and connection with our surroundings is weakened. High-rise apartments, department stores, supermarkets and airports take on a global uniformity and interchangeability. We merely use them for their goods or services so property becomes another commodity to be used in the service of profit.

The solutions to the biodiversity crisis — decreasing consumption and numbers of our species, reducing the inequities between nations, protecting wildlife habitat, eliminating pollution — demand a radical change of attitude to our relationship with the land and other species. Stanford University ecologist Paul Ehrlich believes that "A quasi-religious transformation leading to the appreciation of diversity for its own sake may be required to save other organisms and ourselves."

The seeds of that "quasi-religious" change can be found in the spiritual connection that native people have with their home. There are faint echoes of that attachment in non-native society. It's there in the Saskatchewan farmer who refuses to follow the bank's advice and sell his farm because his father and grandfather farmed that land and he intends to pass it on to his children. This makes no economic sense, but the spiritual meaning of the farm transcends money.

In Newfoundland, the tiny clusters of people who continue to remain in the isolated coastal outports where their ancestors lived and fished for centuries inform us of the reality and power of other values. This is not foolish nostalgia but a hint of a new spiritual relationship with our homes that is of far greater value than anything economic.